FORCES AND
MOMENTS - 1

It is the first book in a large and special series of books, dedicated to motorsport in general; it will cover aerodynamics, suspension, engines, dynamics, etc. ..... Everything you need to learn how to design a full car.

The aim of this series is also to say that I would like to teach again in a university.

I hope that this series will be a success and that I will be able to transmit all my knowledge and all my experience.

@TimoteoBriet

Forces and Momentums

The downforce and drag, are the two forces more important in a race car; booths produce all forces and moments in a car and the full dynamic in track.

	DOWNFORCE	DRAG
	% Total	% Total
Front wing	30	25
Front Wheel	1	12
Chassis	-10	11
Floor and Diffuser	60	15
Rear Wheel	5	18
Rear Wing	27	30

This table is general, but very important to know it.

DOWNFORCE / LIFT

The generation of downforce is one of the most important areas in competition: it is directly related with tire grip; but why is so important?

Let's see, suppose we have a car with a certain weight; we will calculate the maximum cornering speed, with and without downforce ("μ" is the friction coefficient of the tire):

Weight = mass + mass (aerodynamic)   mass = mass
(initial – stop – car)
mass (aerodynamic) = downforce = L   g = gravity
acceleration
ay = lateral acceleration   friction force = μ · weight

$$centripetal\ force = mass \cdot \frac{velocity^2}{curve\ radious}$$

$$friction\ force = centripetal\ force$$

$$velocity = \sqrt{\mu \cdot weight \cdot \frac{curve\ radious}{mass}}$$

$$mass \cdot a_y = \mu(mass \cdot g + L)$$

$$a_y = \mu\left(g + \frac{L}{mass}\right)$$

Suppose: L = X (mass * g) / X > 1
$$a_y = \mu g(1 + X)$$
If   X=1, there are 2g!!!!

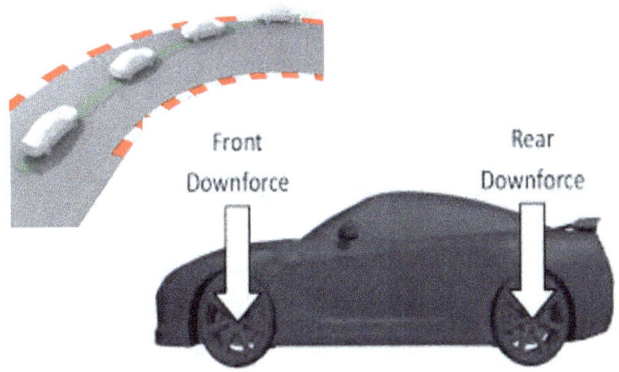

Front
Downforce

Rear
Downforce

In "weight", we should include the weight of the car + the extra weight that the downforce generated; cornering speed will rise with more downforce (it is an "extra" and artificial weight):

- More downforce, higher cornering speed.
- Higher friction coefficient, higher cornering speed.
- Less mass, higher cornering speed; this value will be limited because the car has to stick to the ground, there must be weight.

It is a simple but straightforward expression to obtain approximations: it allows us to know what are the values that optimize the car. We will use it when simulating a lap time.

We know that over the years, the increase in maximum cornering speed (or lateral acceleration) has been due to the improvement in aerodynamics and increased downforce:

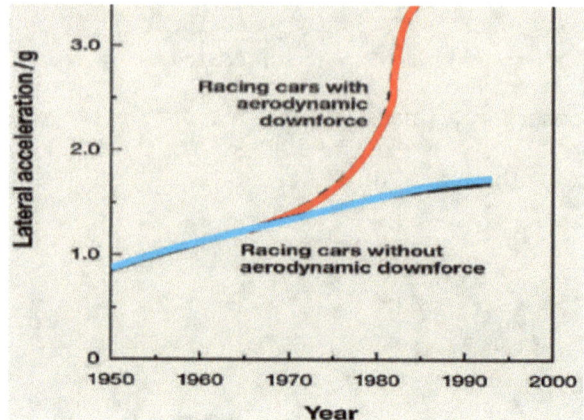

HDF, MDF and LDF correspond to High Downforce, Medium and Low; we can appreciate the difference between speed and time depending on the corner radius and the amount of downforce applied.

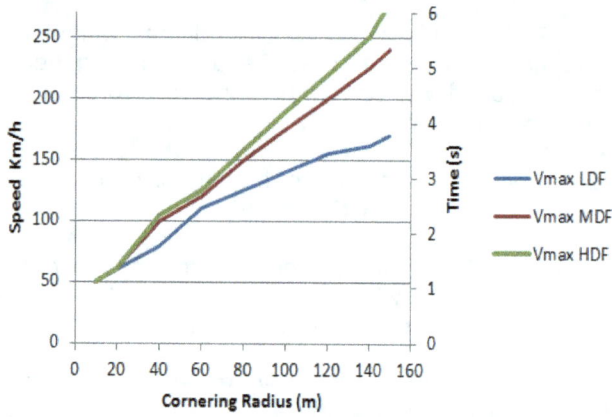

The generation of downforce, has other advantages, such as a reduction in braking time. This occurs because more downforce implies a greater capability to generate braking force by the tire before slip.

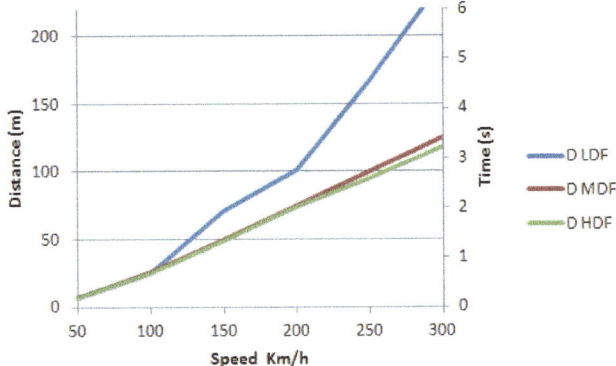

We can see in the following graph an evolution in time of the lap time reduction.

We can see how during the first days the developments in tire and chassis technology where very important. However, a mayor leap forward came with the introduction of wings around 1965, after this date the evolution in aerodynamics has come with an important reduction in lap times.

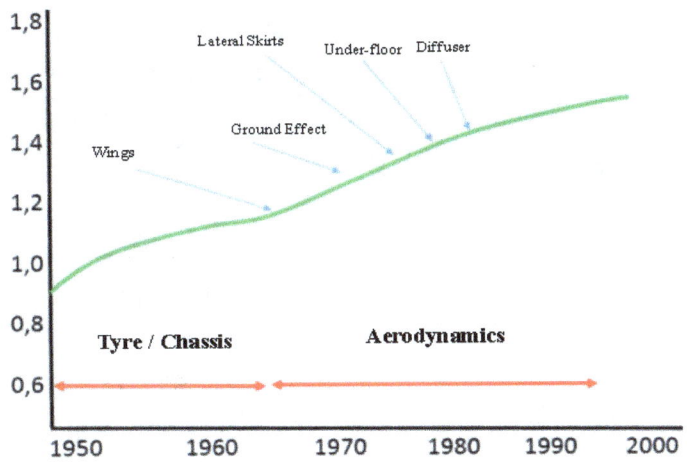

I like to explain the benefits of downforce, as an artificial weight applied on the wheels:

We know that the more weight the tire can bear the greater the friction coefficient will be (with some limits), and therefore, the contact on the ground will be greater.

How downforce occurs?

An air flow generates a force around a surface, by direct impact of the air molecules on the plate; each molecule, although small, has a mass that generates an impact force:

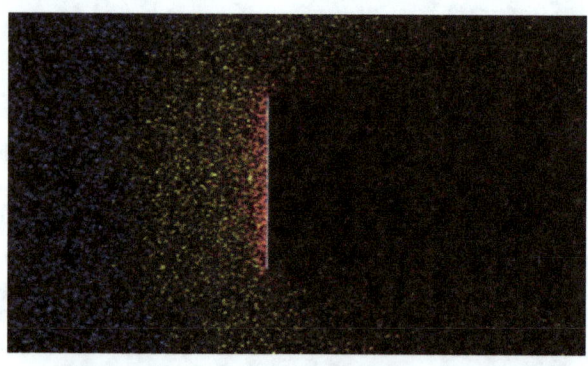

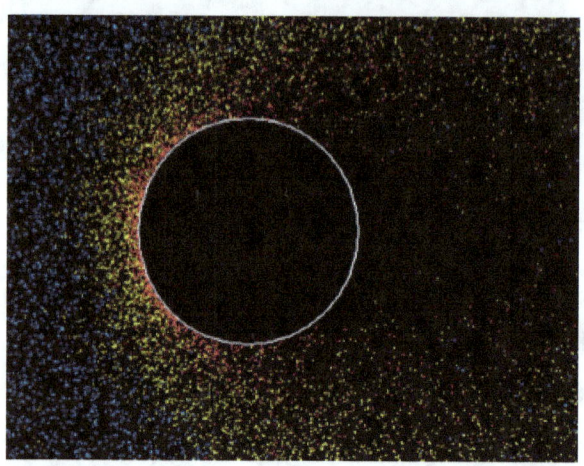

If "n" is the number of molecules colliding with an area "A" and "V" is the velocity of each molecule, the force on the plate area "A" is (where "m" is the mass of each molecule):

$$F = nAmV^2$$

In this expression, no viscosity effects are considered.

Another way of calculating the total downforce could be the following:

Calculate the vertical component of the momentum of each molecule of air "deflected" by the car; the sum of these vector components are equivalent to the total downforce of the car.

It is another way of looking at aerodynamics, considering the air as a set of solid particles.

Given an airfoil, the airflow produces a pressure difference between the top and the bottom sides; in this case, there is less pressure above, thus the wing is sucked upwards; this force is much greater than the one generated by the bottom side. As can be seen in the picture, streamlines above the airfoil are much closer than below the airfoil, indicating that the velocity is greater and the pressure is lower over the foil. This pressure difference produces Lift.

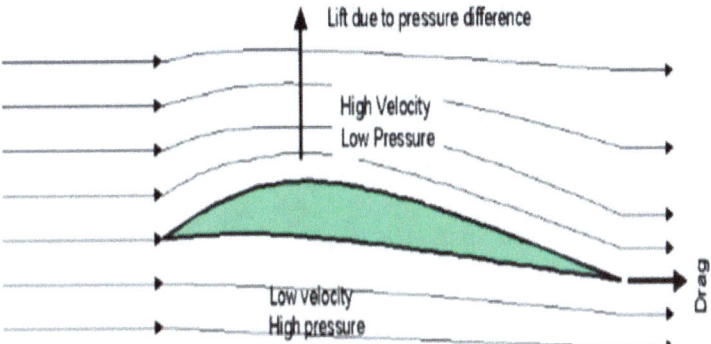

In fact, the up part of wing, have low downforce than down part; for example:

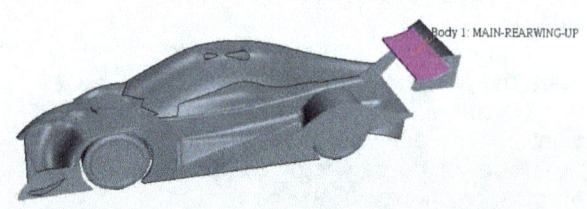

Body 1: MAIN-REARWING-UP

Half Car		
Parts	Newtons	Kg
CABIN	212	21,63265
WHEEL FRONT	98	10
WHEEL REAR	36	3,673469
FLAP REAR WING DOWN	-24	-2,44898
FRONT WING DOWN	-31	-3,16327
DIFFUSER VERTICAL	-55	-5,61224
FLAP REAR WING UP	-58	-5,91837
FRONT WING UP	-196	-20
MAIN REAR WING UP	-309	-31,5306
DIFFUSER	-595	-60,7143
GROUND	-633	-64,5918
MAIN REAR WING DOWN	-710	-72,449

72 N against 31 N i i i i

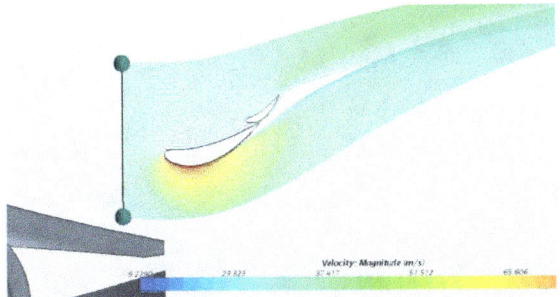

To quantify all the dynamic effects of airflow around a surface, we use the following expression, where "Cl" is the lift coefficient, "A" frontal area, "ρ" density and "V" speed:

$$L = \frac{1}{2}\rho A V^2 Cl$$

Downforce is defined as the force vertically downward and is the vector resultant of all the forces generated on the object.

The quadratic term of the "V" is revealing and very important; fixed at 100 km/h for example, a car can generate 200kg of downforce; we think that is enough, but at 200 km/h, the same car would generate 800 kg of downforce (200 km/h / 100 km/h = 2, 22 = 4, 4 * 200 kg = 800 kg).

The downforce is based on the principle of action-reaction, and it is important to be considered:

If we were able to calculate the mass of air deflected upwards by the car, this mass would be the downforce of the car;

More air going up, means "more car" is going down. We can know the force generated on an angled flat plate, using the Navier Stokes equations; for this, first we calculate the flow on a given flat surface:

Let's focus on the example discussed below:

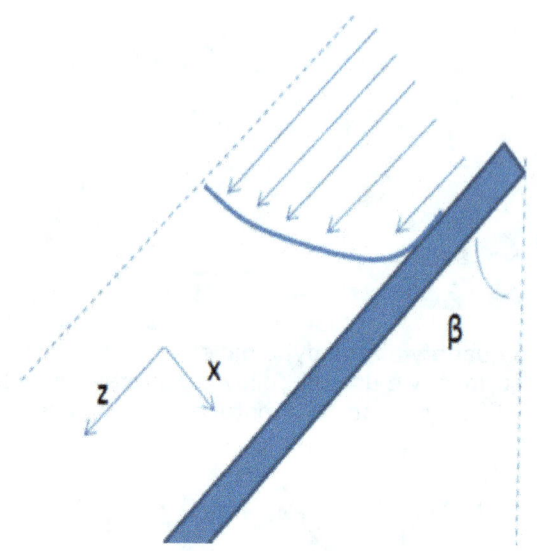

"w" is the size of the surface and "H" (as above)
is the height flow rate's layer you want to know :

$$v_z(x) = \frac{\rho g \cos \beta}{2\mu}(H^2 - x^2)$$

$$A = \frac{\rho g \cos \beta}{2\mu}$$

$$Q = \int_0^w \int_0^H v_z(x)\,dxdy$$

$$= \int_0^w \int_0^H A(H^2 - x^2)\,dxdy = \int_0^H (A(H^2$$

$$- x^2)y)\,dx = Aw\left(H^3 - \frac{H^3}{3}\right)$$

$$= \frac{\rho g \cos \beta}{\mu}w\frac{H^3}{3}$$

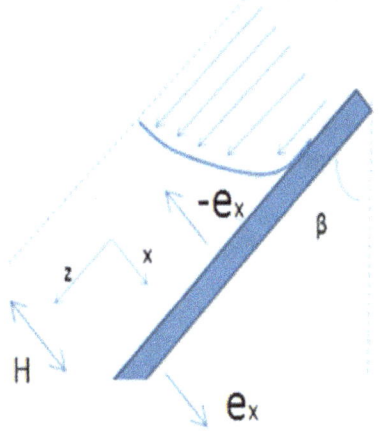

"L" is the length in "y".

$$v_z(x) = \frac{\rho g \cos(\beta)}{2\mu}(H^2 - x^2)$$

$$F = \int_0^L \int_0^H stress(plate)dydz$$

$$Force = area * stress$$

$$stress(plate-in-z-direction) \rightarrow \tau_{xz} = -\mu \frac{dv_z}{dx}$$

$$F = \int_0^L \int_0^w -\tau_{xz}(surface)dydz = \int_0^L \int_0^w -\tau_{xz}(x=H)dydz$$

$$\frac{dv_z}{dx} = \frac{\rho g \cos(\beta)}{2\mu}(-2x)$$

$$\tau_{xz} = -\mu\left(-\frac{\rho g \, \cos(\beta)}{\mu}\right) = \rho g \, \cos(\beta) x$$

$$F = \int_0^L \int_0^W (\rho g \, \cos(\beta) x)(x = H) \, dy \, dz =$$

$$\int_0^L \int_0^W -\rho g \, \cos(\beta) H \, dy \, dz = -\rho g \, \cos(\beta) H \int_0^L \int_0^W dy \, dz =$$

$$-\rho g \, \cos(\beta) HWL$$

We rely on the air to push the car downwards; hence at low temperature air is "heavier" and thus you need to move up less flow t generate the same downforce. At low temperatures, aerodynamically, the car runs better (higher density). We define the angle of attack of a profile as the angle between the horizontal and the profile's chord; also another's definitions:

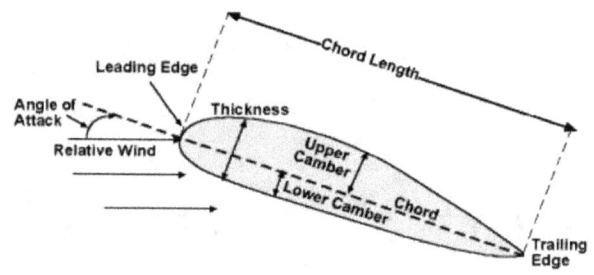

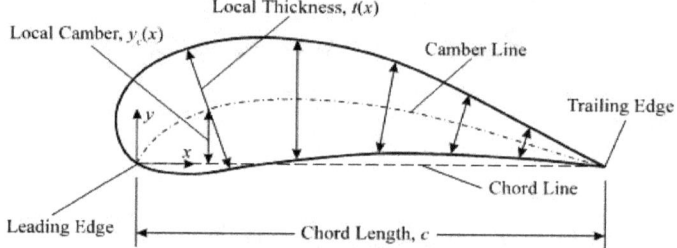

In this moment, is necessary know the influence of parameters (thickness and camber), in a downforce and drag):

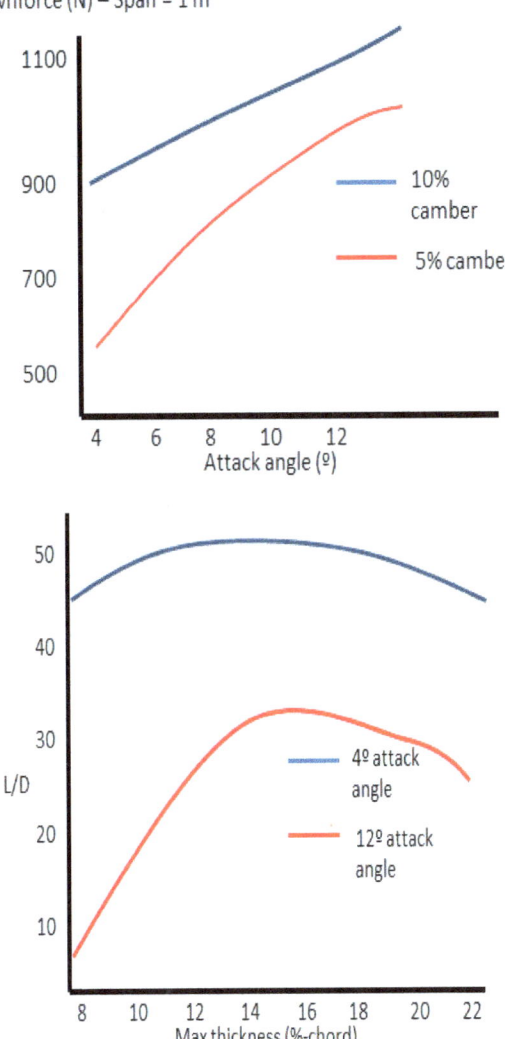

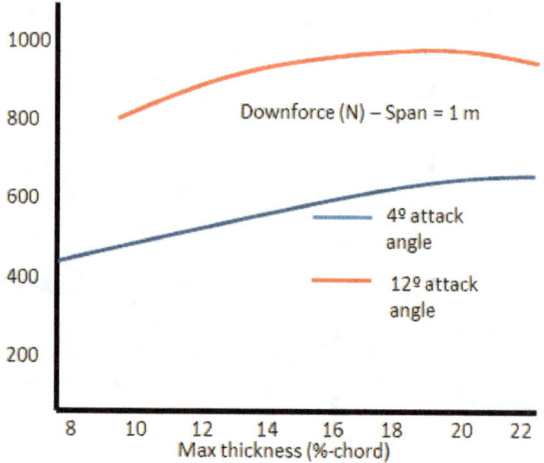

The greater the angle of attack the higher the lift will be; This statement is true up to a certain angle called "stall angle"; Above this angle lift will not only not increase, but it will drop rapidly. This is mainly due to the detachment of the flow:

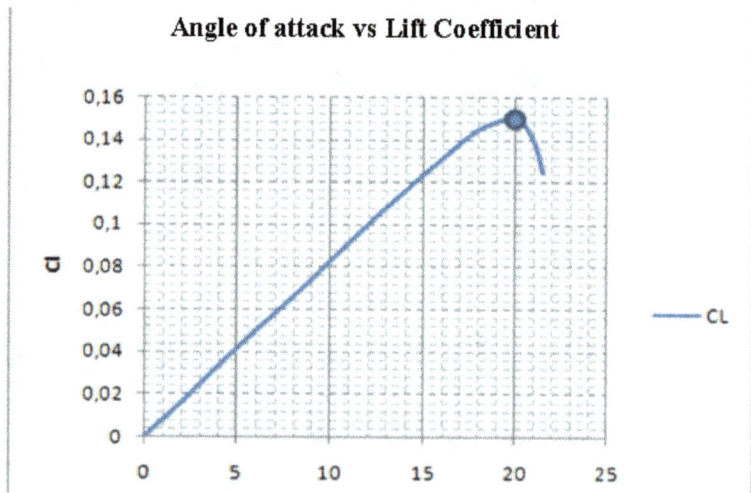

How to measure the angle of attack of a wing? Theoretically it is very difficult to know this angle of attack; instead of using the "theoretical" angle, it is often used another angle which is often easier to measure.

The process is simple: we place a protractor on the wing and we measure the angle:

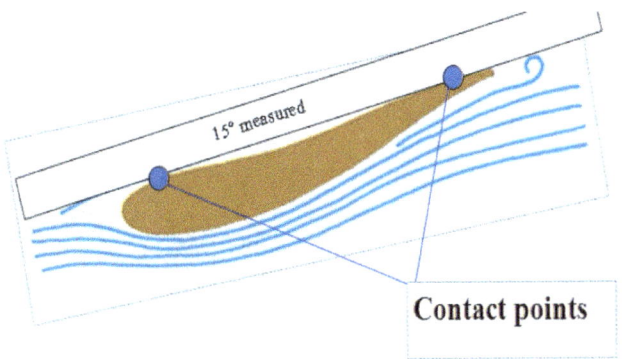

15° measured

Contact points

But there is a one special case that will be explained next. The rest of the chapter will be described by Alejandro Murillo Juliá, an Aeronautical Engineer graduated in the Technical University of Madrid (UPM). His project (summary):

"The project aims to make the research of the aerodynamic performance of the NACA 63-415 profile. To do so, a process of design, manufacturing and assembly of the models is developed in order to make subsequent wind tunnel testing at a Reynolds number of 500.000. Based on the pressure readings, measured in the AB6 wind tunnel, several characteristic curves that describe the profile's behavior will be obtained over the entire range of angles of attack, from 0 to 360°. Once obtained, the aerodynamic coefficients will be compared with CFD simulations (Star-CCM+ and Fluent), panels codes (XFOIL and JavaFoil) and technical references, in order to verify the validity and accuracy of the results."

The next data, graphics and results have been obtained by the wind tunnel tests of the NACA 63-415 profile. For this profile, the lift curve will be represented, showing how the lift coefficient changes with the angle of attack:

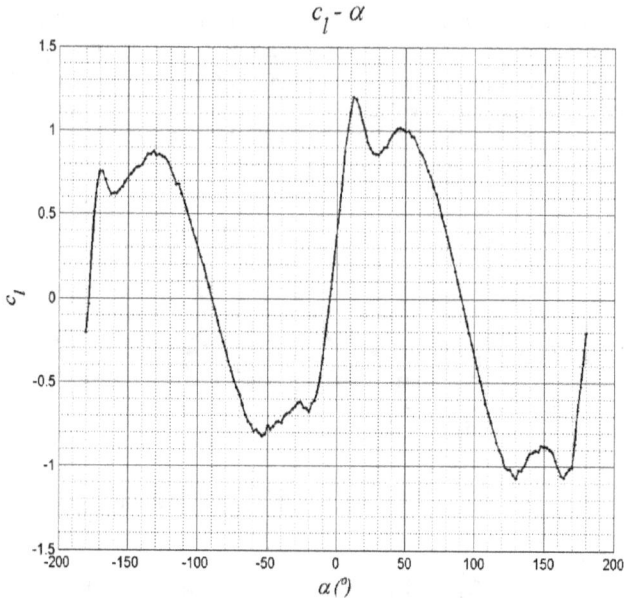

c_l - α

Graphics developed by Alejandro Murillo Juliá, an Aeronautical Engineer, as part of his final degree project; values obtained by wind tunnel testing in the IDR (Ignacio Da Riva Institute). Alejandro was student of COURSE Aero and CFD in 2015 August (Timoteo Briet as professor).

It can be seen that the results obtained when the profile rotates in both directions are very similar. Although the peak reached when the profile rotates in one direction is different to the peak reached when it rotates in the other direction, due to the profile's camber line. Also, we can see the same effect in the drag curve:

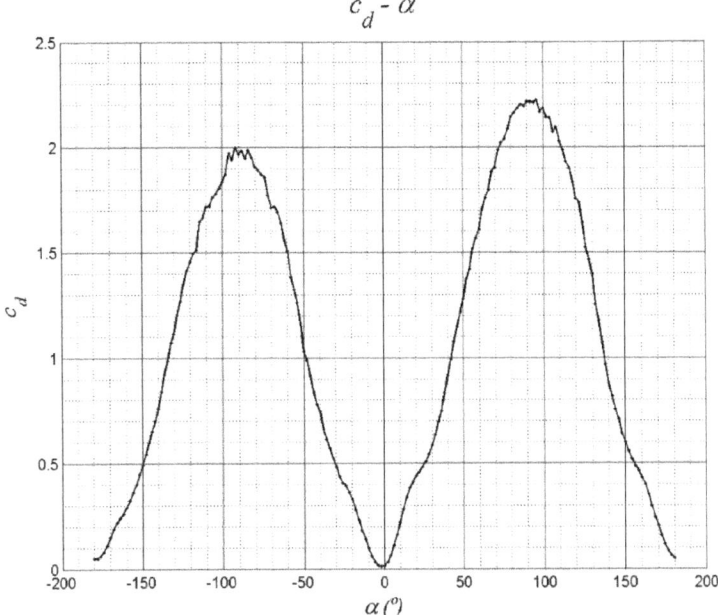

$$c_d - \alpha$$

The linear region of the curve can be clearly identified in the central part of the graphic, where the C_l rises constantly when the angle of attack rises, until the profile reaches the stall angle. At this moment the lift curve drops in two different sections. This phenomenon is known as "**double stall**". It is shown how, when the maximum is surpassed, and the angle of attack continues rising, the lift curve suffers a sharp drop and, then, the lift coefficient rise again, before reaching a second peak which precedes to a dramatic drop in the lift coefficient. This second peak is due to the recirculation bubble or reattachment vortex: once the stall angle has been surpassed, the detachment of the flow will take place and it will generate a low pressure vortex that, when the angle of attack continues rising, will allow the reattachment of the flow. Then, the effect of the separation bubble will result in a raise in the lift coefficient after the stall point. When the angle of attack rise, the bubble increases its size until it reaches the trailing edge, when the flow can't be reattached as turbulent flow. In this case, at $\alpha = 50°$, the pressure gradient will become so adverse that the air will not be able to reattach again and the flow will separate, generating a wide wake and the lift will fall completely.

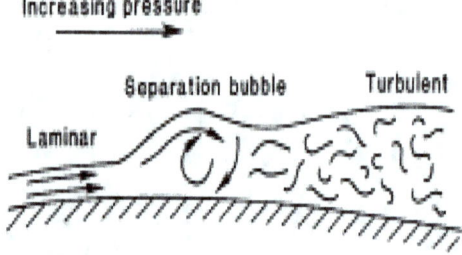

Separation bubble and reattachment of the flow at Re=500.000

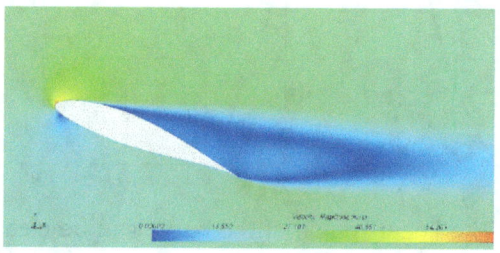

Scalar field of velocity magnitude of the NACA 63-415 profile at α=20º, obtained by CFD simulations in STAR-CCM+ developed by Alejandro Murillo.

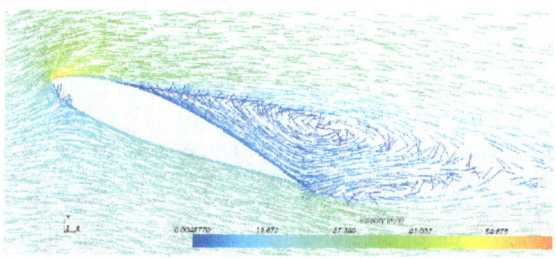

Vector field of velocity of a CFD simulation at $\alpha = 20°$, when the vortex and the detachment of the flow can be clearly seen.

Streamlines of the simulation of a CFD simulation at $\alpha = 20°$, when the vortex and the detachment of the flow can be clearly seen.

When the angle of attack is, more or less, set up at $\alpha = 90°$ the lift becomes negative due to its geometrical situation, the camber line will be inverted. When the angle of 180° is reached, the lift cycle behavior will repeated itself again and the differences in the peaks of maximum drag and lift will be explained by the greater or lesser curvature of the mean line.

If the linear region of the lift curve is expanded, for little angle of attacks, some phenomena can be observed:

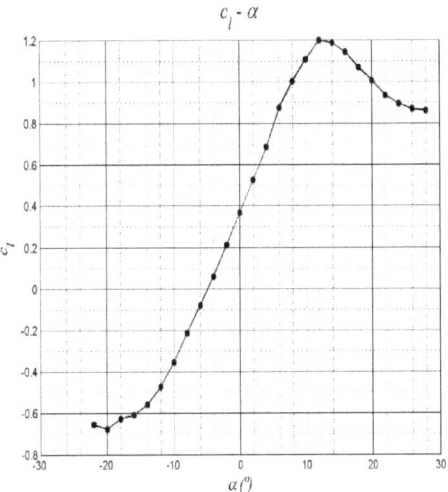

In the linear region, the lift coefficient at an angle of attack of zero (where the lift is only due to the distribution of thickness and the camber of the profile) obtained is C_{l0}=0,3688, the maximum lift coefficient is $C_{l\,max}$=1,1992, reached at the stall point, placed at α_s=12° and a slope of the lift curve of $C_{l\alpha}$=0,0803 (1/°).

Next, the results in the linear zone of the lift curve obtained in the wind tunnel tests will be compared with different techniques, including CFD simulators, linear potential theory and catalogs of wind tunnel testing results.

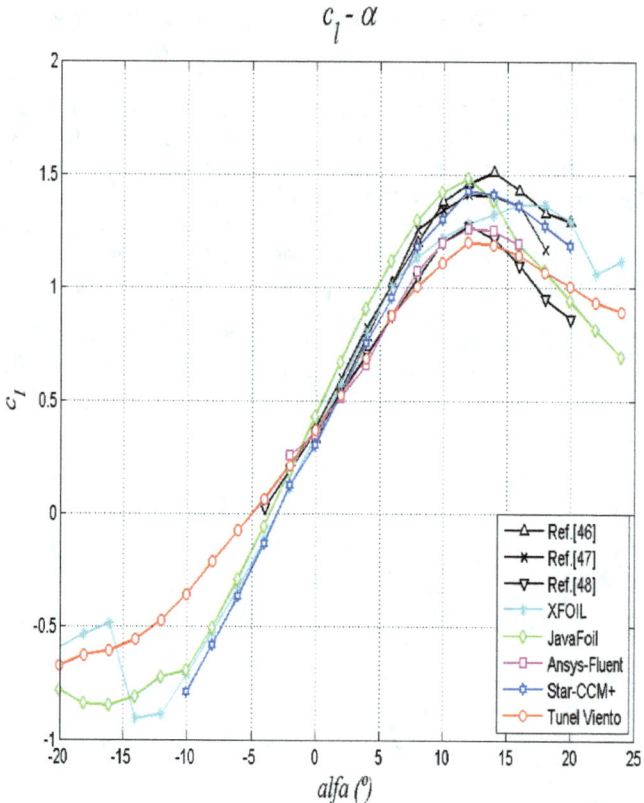

$c_l - \alpha$

In the following table, the characteristic values of the lift curve in the linear region are attached, with the relative errors of the wind tunnel tests, referring to the each method to provide the information required to study accurately and to determine the validity of the experimental results

$$\varepsilon_i(\%) = \frac{Wind\ Tunnel\ Result - Method\ "i"\ Result}{Method\ "i"\ Result} \cdot 100$$

Method	Re	C_{l0}	ε_{l0}	$C_{l\,max}$	$\varepsilon_{l\,max}$	α_s (°)	$\varepsilon_{\alpha s}$	$C_{l\alpha}$ (1/°)	$\varepsilon_{l\alpha}$
Wind Tunnel	$0,5\cdot10^6$	0,37		1,20		12		0,08	
Star-CCM+	$0,5\cdot10^6$	0,30	22%	1,42	-15%	12	0%	0,11	-27%
Ansys-Fluent	$0,5\cdot10^6$	0,35	6%	1,26	-5%	12	0%	0,07	11%
JavaFoil	$0,5\cdot10^6$	0,43	-14%	1,48	-19%	12	0%	0,12	-32%
XFOIL	$0,5\cdot10^6$	0,35	6%	1,36	-12%	18	-33%	0,11	-29%
Ref. [48]	$1,6\cdot10^6$	0,37	0%	1,27	-6%	13	-8%	0,09	-6%
Ref. [47]	$3\cdot10^6$	0,38	-3%	1,41	-15%	13	-8%	0,11	-28%
Ref. [46]	$3\cdot10^5$	0,33	12%	1,51	-21%	14	-14%	0,12	-30%

It can be observed that there are some differences between the results obtained by the different methods. Next, this origin of these differences will be explained shortly:

- In the first place, to understand the differences between the results obtained in the AB6 wind tunnel and the results of the reference catalogs, due to the test conditions set up a different Reynolds Number, it is necessary to know the influence of this parameter in the aerodynamic performance of a profile. Next, two images can be found attached, showing the effect of this parameter. It can be seen that, when the Reynolds decreases, the maximum lift coefficient suffers a sharp drop.

- On the other hand, the results obtained by panel codes or CFD simulations, at the same Reynolds Number, are more similar to the values obtained at processing the wind tunnel measures, especially at an angle of attack of zero. However, the differences at the maximum lift coefficient and the slope of the curve are associated to the roughness of the wind tunnel test model and the turbulence model chosen at the CFD simulations (K-epsilon).

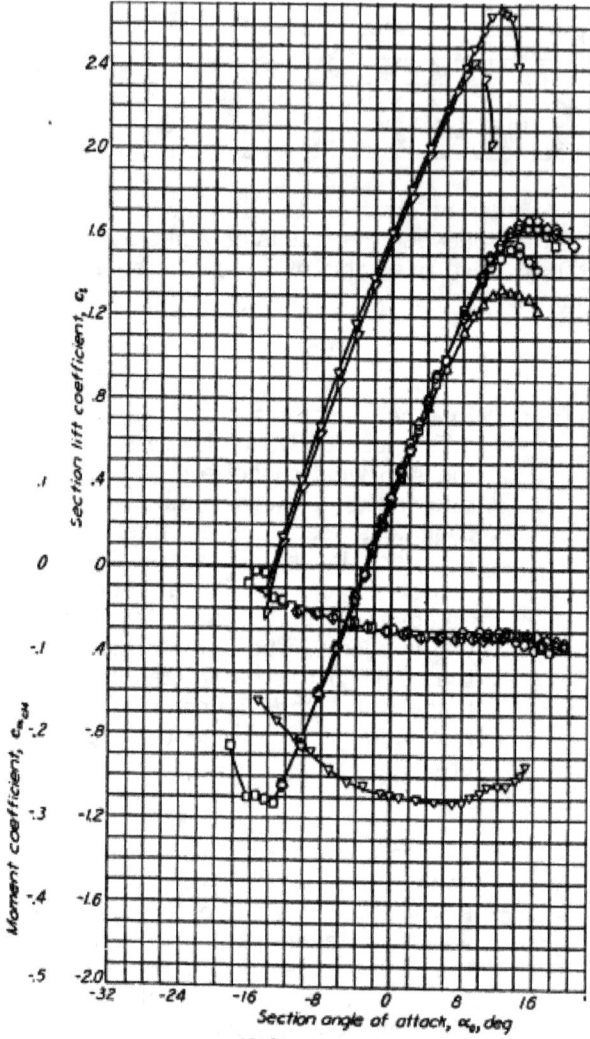

NACA 63₁-415 Wing Section

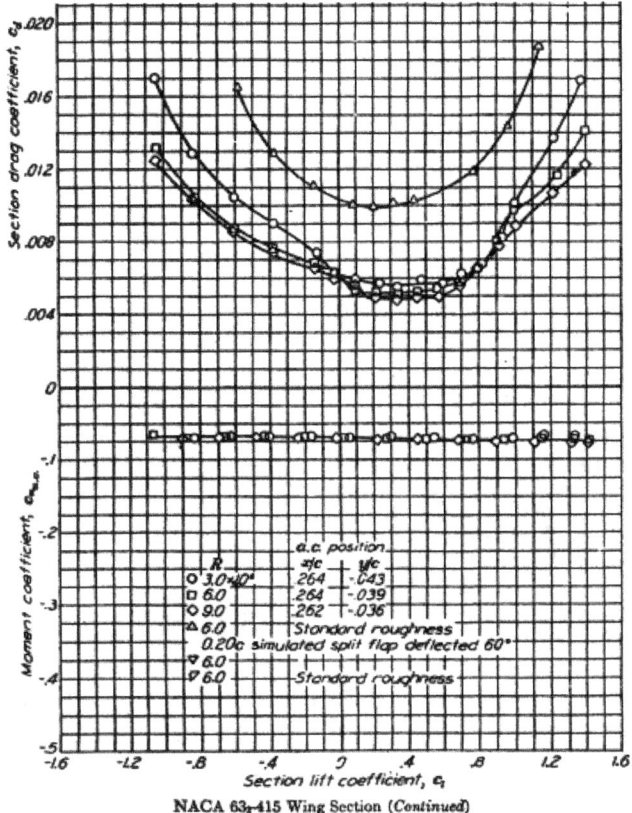

	R	a.c position	
		x/c	y/c
○	3.0×10⁶	.264	-.043
□	6.0	.264	-.039
◇	9.0	.262	-.036
△	6.0	Standard roughness	

0.20c simulated split flap deflected 60°

| ▽ | 6.0 | | |
| ▽ | 6.0 | Standard roughness | |

Section lift coefficient, c_l

NACA 63₁-415 Wing Section (Continued)

The variations of the lift coefficient versus angle of attack can be very significant; in the graph below you can see different values depending on "h/c", where "h" is the height of a NACA 0012 profile with respect to the ground and "c" is the chord of the profile. As you can see, the height of the airfoil is also a very important characteristic:

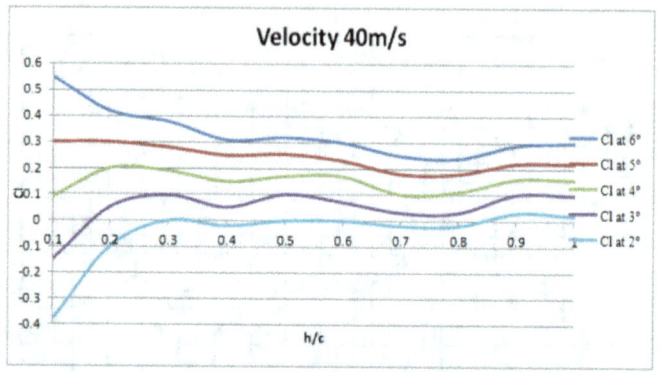

Example: How to calculate the lift coefficient of a Boeing 747 wing:

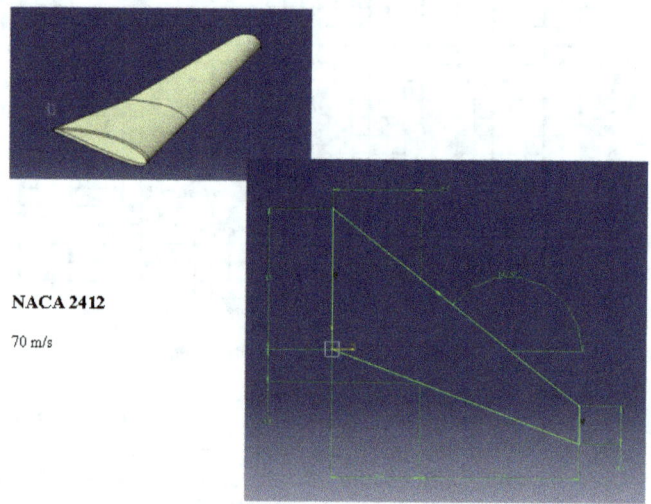

NACA 2412

70 m/s

In the first length: $0 < x < 9.7$ the chord expression is $c(x) = 15 - 0.4062x$ so downforce is:

$$L_1 = \frac{1}{2}\rho v_\infty^2 \int_0^{9.7} (15 - 0.4062x)\, dx$$

$$= \frac{1}{2}\rho v_\infty^2 \left[15x - \frac{0.4062x^2}{2} \right]_0^{9.7} = \frac{1}{2} \cdot 1.225$$

$$\cdot\, 70^2 \cdot 126.39$$

So:
$$L_1 = 379330 \cdot Cl$$

In the second length $9.7 < x < 27.2$ the chord expression is:

$$c(x) = 11.06 - 0.3977(x - 9.7) = 14.9177 - 0.3977x$$

$$L_2 = \frac{1}{2}\rho v_\infty^2 \int_{9.7}^{27.2} (14.9177 - 0.3977x)\,dx$$

$$= \frac{1}{2}\rho v_\infty^2 \left[14.9177x - \frac{0.3977x^2}{2} \right]_{9.7}^{27.2} = \frac{1}{2}$$

$$\cdot 1.225 \cdot 70^2 \cdot 132.6$$

So:
$$L_2 = 398000 \cdot Cl$$

Finally, assuming constant Cl, the total downforce is the sum of both:

$$L = L_1 + L_2 = 777330\ Cl$$

Another example: Calculate the lift of helicopter wing:

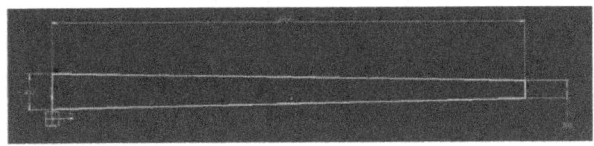

NACA 0012 at 200 rpm.   Chord 40 cm and 20 cm and wingspan 6 m.

$$L = \frac{1}{2}\rho \int_0^l Cl(x)c(x)v_\infty^2\,dx$$

$$v_\infty^2(x) = (\omega \cdot x)^2$$

$$c(x) = c_1 - \frac{c_1 - c_2}{l}x$$

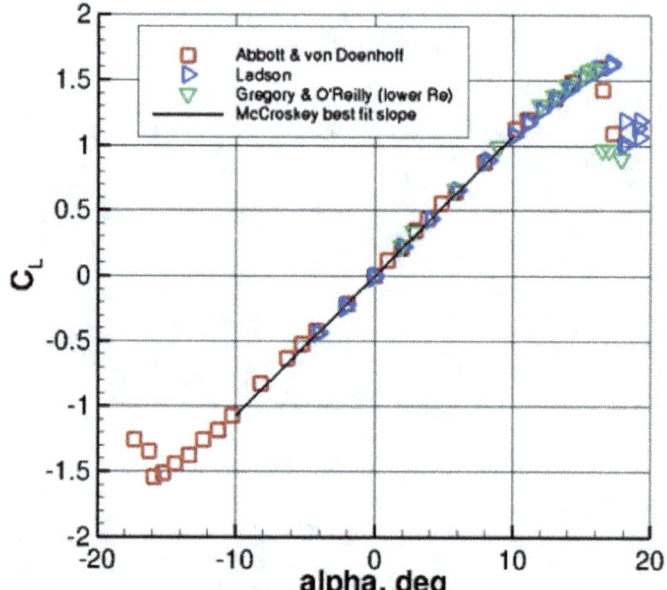

There is torsion between 2º and 10º:

$$Cl(x) = Cl(2^\circ) + \frac{Cl(10^\circ) - Cl(2^\circ)}{l}x$$

$$v_\infty^2(x) = 438.65x^2$$

$$c(x) = 0.4 - 0.0333x$$

$$Cl(x) = 0.18 + 0.145x$$

$$L = \frac{1}{2} \cdot 1.225 \int_0^6 468.35x^2 \cdot (0.4 - 0.0333x) \cdot (0.18 + 0.145x)\,dx$$

$$L = 3902\ N = 398.2\ kg$$

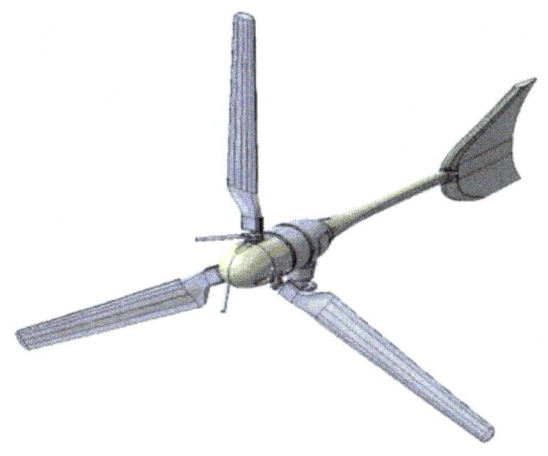

INCREASING DOWNFORCE OF A WING

Let's see how we can increase downforce in a racecar wing; there are 3 ways:

1) Multi-element wings.
2) Gurney Flap.
3) Turbulators.
4) Others.

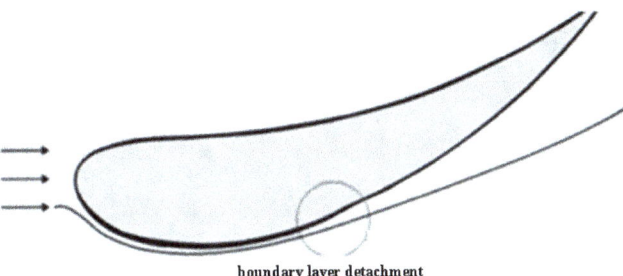

boundary layer detachment

1) Multi-element wings:

If we increase the angle of attack of this profile, there will be a moment when the flow will detach from the surface, reducing downforce and increasing drag; to prevent this from happening, we placed another profile on top, creating a small gap between the two of them; this gap, accelerates the flow reducing the pressure preventing detachment. This means we can now give a greater angle of attack to our wing to increase downforce:

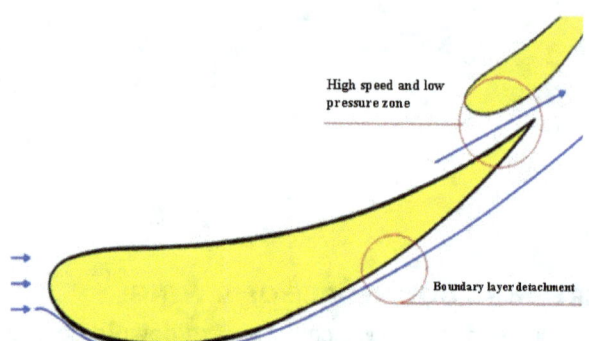

We are increasing the stall angle of the profile; if we do add more and more profile elements more downforce will be generated. However, we should check the rules of each motorsport series because they often restrict the use of flaps up to a certain number. In most series, just 2 are authorized.

We can use CFD to test the position and angle of the different flaps that form the wing-set. A good starting point to further optimize is the following:

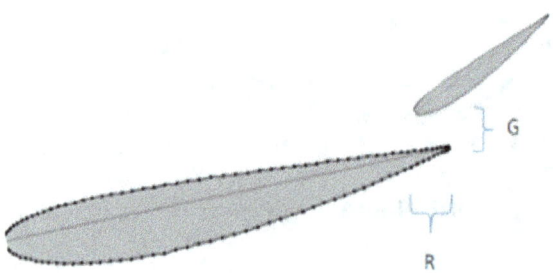

- R = 2% of the sum of the chords (of 2 profiles).

- G = 2.2% of the sum of the chords (in the two profiles) - (2% - 3%).
- The maximum angle of the flap should be around 25º- 33º.

Another rule:

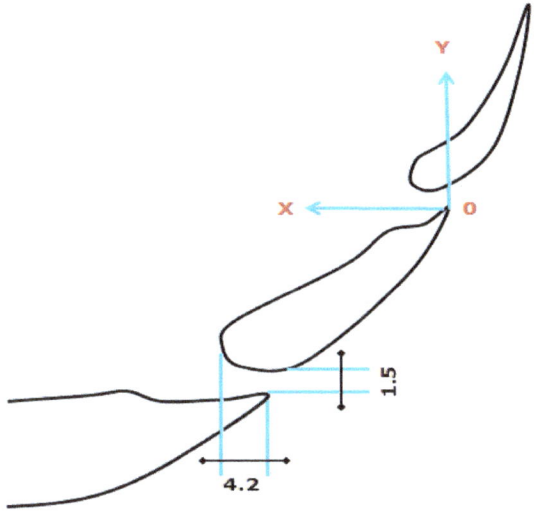

We can go up and up adding new flaps increasing the angle of attack of each new flap (even more than 90º):

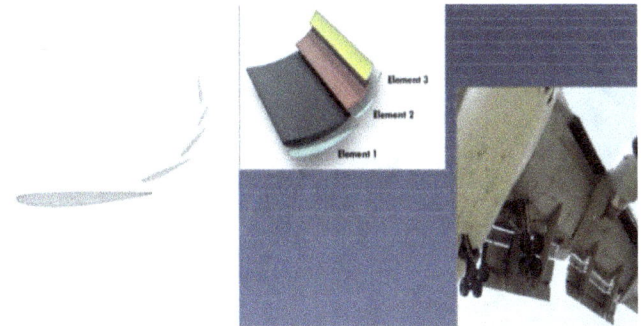

Where is the limit? The limit appears the moment downforce is no longer generated and drag is increased.

Assume the following wing profile: is this a good design?

The answer is no, because the flow is quickly detached from the bottom, given the extreme amount of pressure gradient (geometric gradient); i.e. the air won't have enough energy to travel across all the bottom side. We can get much more downforce with less drag dividing the wing in different flaps:

This last design has greater -to-drag ratio (L/D) meaning it has more downforce and less drag. This term is sometimes referred as aerodynamic efficiency.

In boat wings, is common to see wings with 9 or more parts:

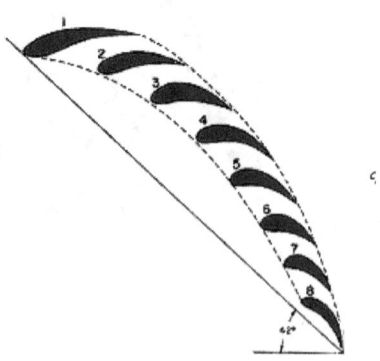

Ia amazing: the lift that is possible to create trough this method, is incredible:

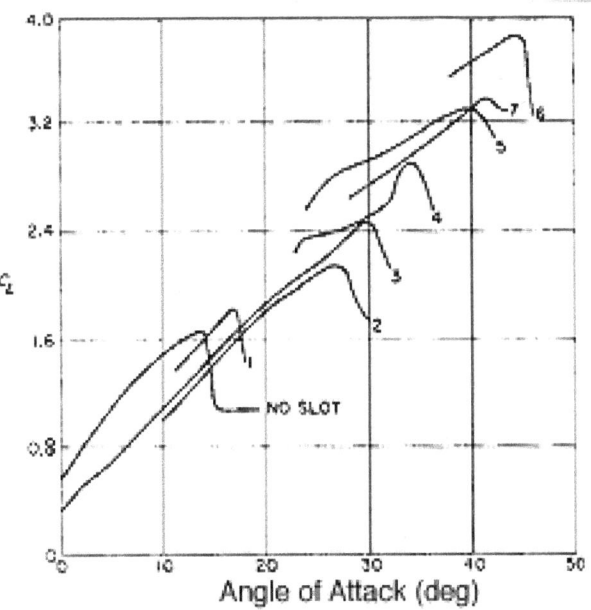

Angle of Attack (deg)

Efforts have been made to avoid the detachment of the flow on a low pressure high speed side of the wing (bottom side on a race car), as the force generated by this side is much greater than the one generated by the upper side.

Boundary layer detachment is dependent on the curvature of the wing, the leading edge and, very importantly, the flow rate. If the speed is high, the phenomenon occurs before, so if we are at a low speed circuit we can choose configurations with fewer flaps.

Thus, in circuits like Monaco and Hungary it is likely to see "simpler" noses than the ones we see in England for example.

Albert fabrega

2) The Gurney flap:

The Gurney flap, is something different and based on a different principle; Let's assume we have a wing, in which we are at the stall angle or even surpassing it; we can place a gurney flap in the trailing edge. This flap is a small plate perpendicular to the wing's chord; It will generate a depression on the suction side increasing pressure on the pressure side, doing this avoids detachment of the boundary layer.

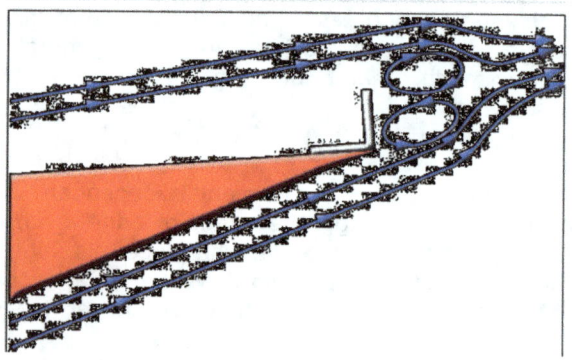

This Gurney flap usually has no more than 2.5 cm and it can be installed almost anywhere: front wing, rear wing, etc...

The Gurney is installed where a relative low pressure is needed, thus sucking air can act to:

- Increase the angle of attack avoiding stall.
- Use the system to channel air to a certain

place.

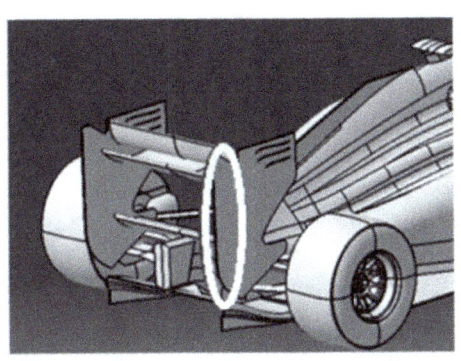

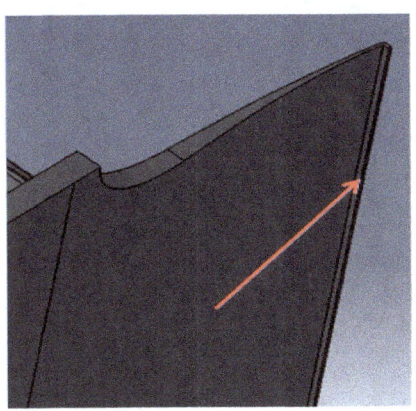

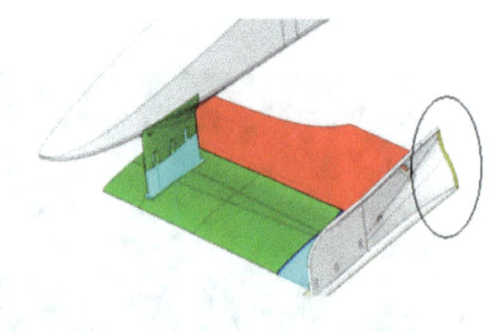

Here we can see the Gurney doing a better extraction of air from the cooling duct, which also serves to improve the diffuser (Albert Fabrega images):

In the image (Albert Fabrega), note the installation of a "circular Gurney"; the goal of this flap is that the nozzle or expansion zone works properly filled with air.

There is a system called "Monkey Seat"; this is a small system introduce in 2014 season that "deflects" the air; depending on the team, there are several design alternatives (Albert Fabrega images):

The monkey seat, is used for a lot goals:

- Generate downforce.
- Reduce drag, filling the low pressure zone in rear car.
- Generate downforce from improving rear wing.

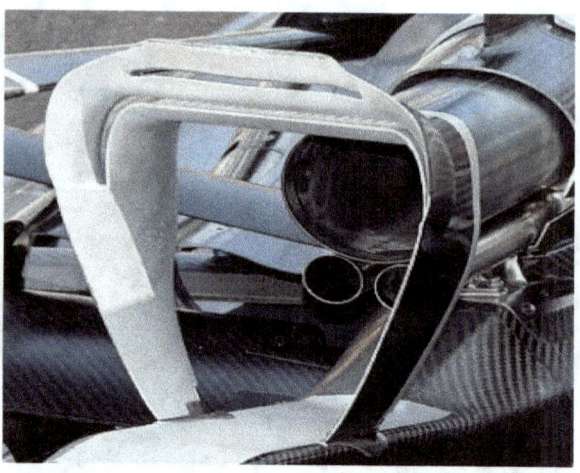

That last goal is very important. The flow, is forced to attached to down rear wing. This makes it possible to generate greater incidence angle for rear wing without stall.

It has always been said that Gurney flaps should be only placed on those wings with an angle of attack close to stall; This is true, but it is not completely accurate; if we have a wing with a small angle of attack and we install a gurney flap, lift will be almost the same, but drag is greatly reduced. The effective geometry for the flow of air will be different more refined, generating less drag.

Virtual Geometry

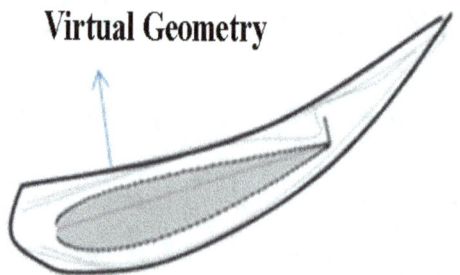

The same applies to the general "virtual shape" of a car; it tends to generate lift:

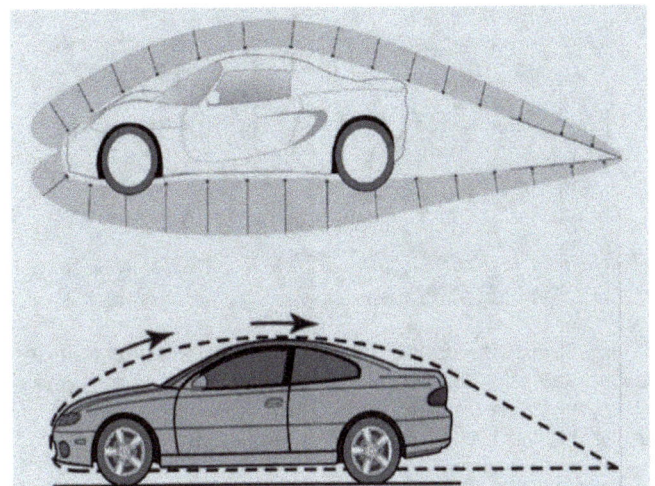

So we can ask ourselves what is the best virtual geometry that generates the least possible drag: People tend to answer that the teardrop shape is the least drag generating shape; however, it is a false answer. In fact the answer to the above question is that the best and the most optimal shape is a sphere "modified"; i.e. the generated virtual geometry:

In fact, when a drop is falling it has a spherical shape. Only in the case of being "hung" takes this form, because of gravity:

If we represent the streamlines around a sphere, we can see the "traditional teardrop" that we all know; it is the best aerodynamic shape in the sense of producing the least drag (Wikipedia images):

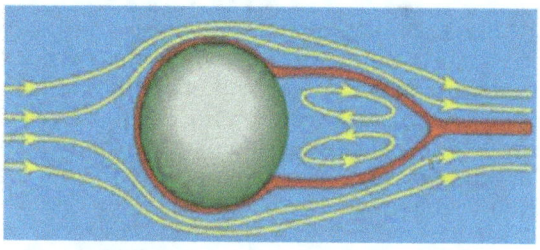

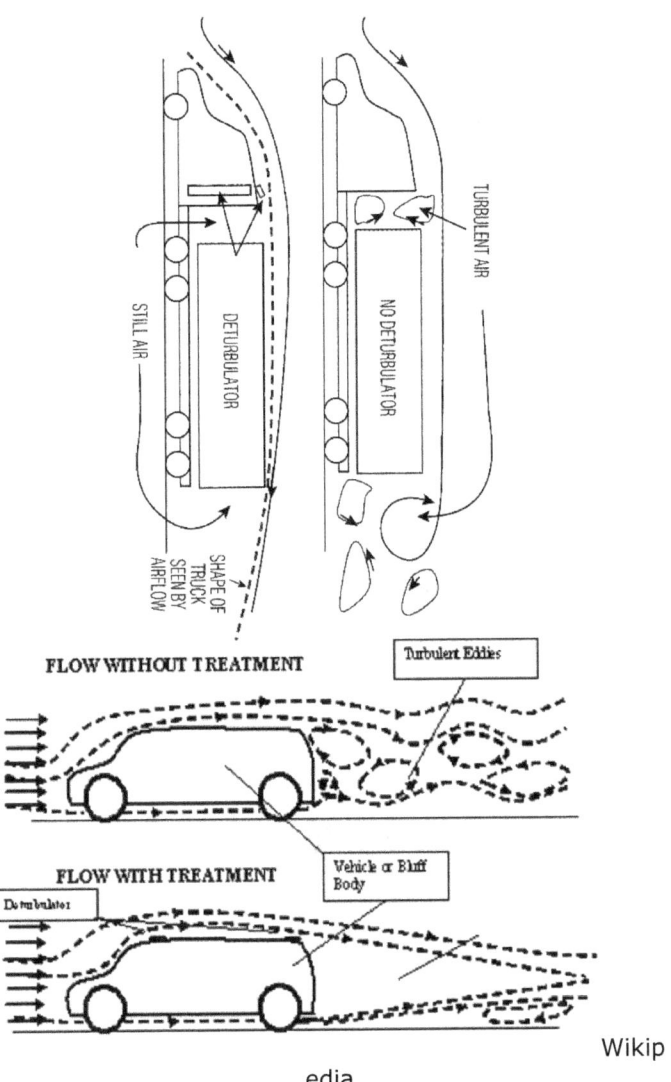

Wikipedia

The shape that any geometry downstream has is much more important than frontal area. It's something that usually, in preliminary designs of cars, is not taken into account, but should be considered; Drag, downforce and therefore the global dynamics of the car, depend largely on how air flows around the object and how it evolves further back.

Another think very important and false...., is the analogy with nature; a lot times we say that the nature is the best inspiration in order to improve or create a shape for one aircraft; for example:

In the first case, we are a bird and the second case, we are a aircraft (side views). Both are more or less the same geometry; so is possible say the commented before.

But no; that is false; why?
The speed in aircraft is big; for this reason (by Reynodls number or theory), the speed for the bird (low speed) in order to have the same geometry, must to be very very big.....

Another think or reason (false idea): the images are by side views only....

It is very important not to set too much on the nature ....

Other sample about low drag: shells:

Being a water drop shape, the shape with less drag means that a reduction of the front section downstream would decrease drag:

We can learn, as always, of nature itself; some species of owls take this "pointy" position to achieve higher speeds (Wikipedia images):

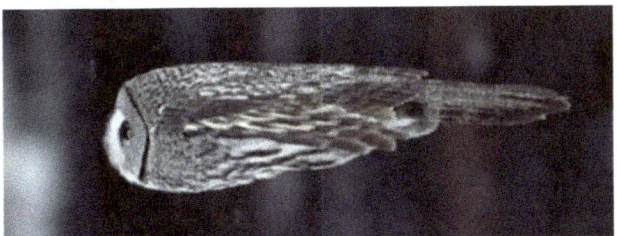

Finally it must be said that the Coanda effect is responsible for circulating the air into the rear properly and without turbulence.

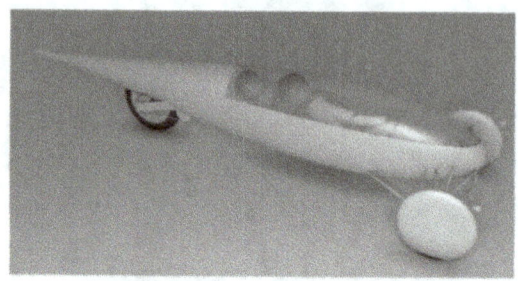

Following the above again, sometimes the Gurney flap is inverted:

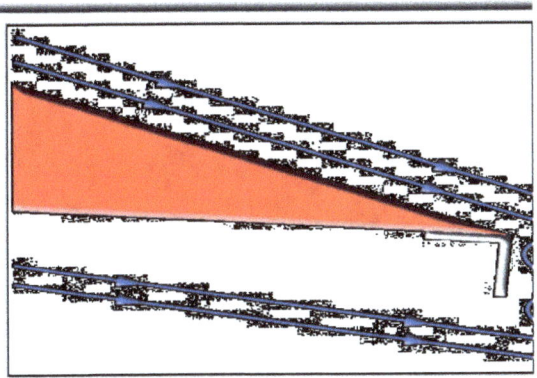

It is effective in reducing the downforce without lowering the angle of attack;

This method was used by McLaren in the 80s.
Moreover, there may be some rare gurneys and more aerodynamic gurneys coming from analyzing birds; such is the case with saw tooth gurneys:

Gurney flap height is typically between 1% and 4% of the wing's chord; we can see the influence that this height has on downforce (lift coefficient):

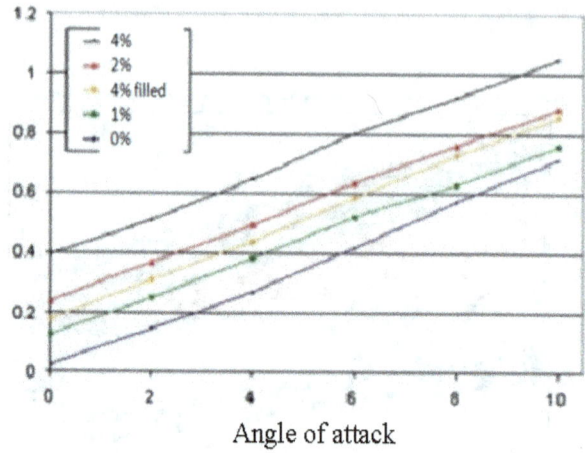

Angle of attack

We know that turbulent flow tends to adhere more and better on angled surfaces; for this reason, if the wing has a considerable impact, the air flow tends to detach; for this reason, we can create turbulence in the flow so it does not detach. These devices are called turbulators and are just that: responsible for making the flow becomes turbulent.

3) Turbulators (also applied to floor ad difusser, and also to reduce drag):

We know that turbulent flow tends to adhere more and better on angled surfaces; for this reason, if the wing has a considerable impact, the air flow tends to detach; for this reason, we can create turbulence in the flow so it does not detach.

These devices are called turbulators and are just that: responsible for making the flow becomes turbulent. So we can achieve higher angles of attack without flow detachment, generating higher values of downforce.

For example, these turbulators on the roof of the Mitsubishi EVO makes the flow turbulent and fix it all over the rear window, reducing the drag and enabling the operation of the aileron.

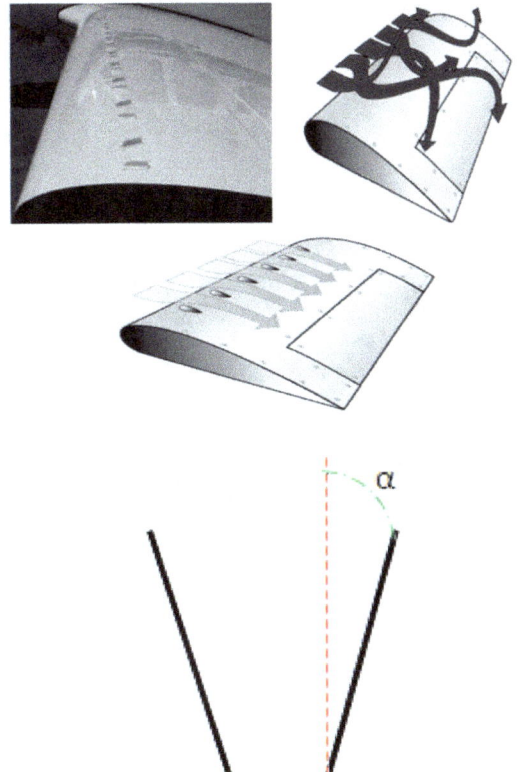

Angle between 12 and 18º, more or less.

We can have vortex in the same rotation or different:

Counter – rotating

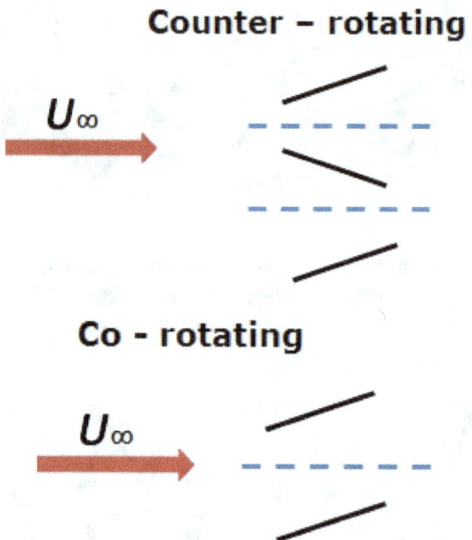

U_∞

Co - rotating

U_∞

Is necessary to study the evolution back of vortex.

In fact, we can see the next example – test; the incidence angle, affect to lift of this flat plate:

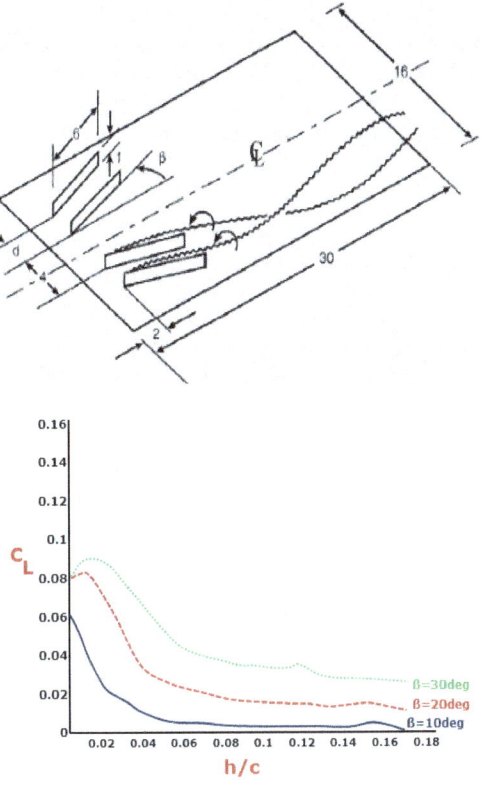

And also, if theses vortex generator are in ground car, the roughness of track, can affect to vortex evolution; that is important study it.

→ *Another function of vortex is:*
To create vortex, in order to help air to go where is needed. Energizer the air flow, is very important, because if the flow not have speed low, is complicate change his direction....

Another vortex generator type (Wikipedia images):

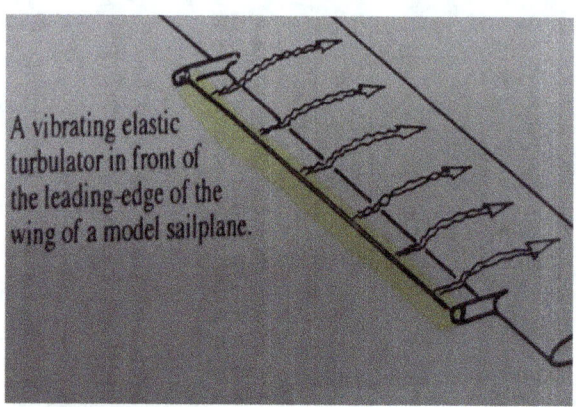

A vibrating elastic turbulator in front of the leading-edge of the wing of a model sailplane.

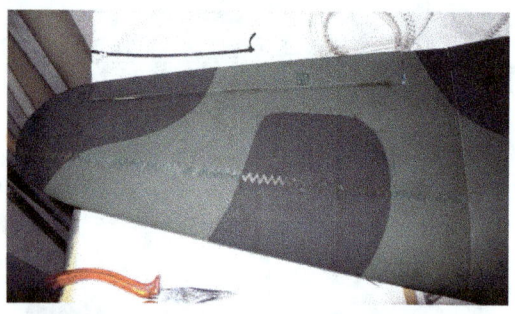

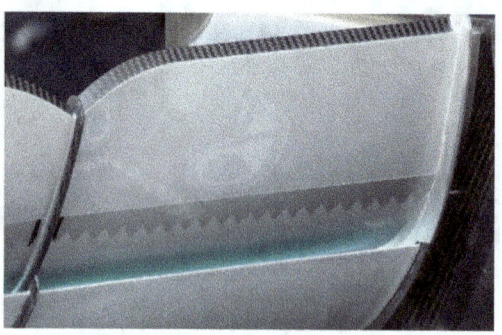

These two last images: vortex generators in order to improve the incidence angle.

4) Other systems:

We have left to the end this type of profile, for its peculiarity; its operation is simple: at each "step" a depression is created that sucks the air to keep the flow attached to the surface; simple and brilliant:

Family called KFM profiles.

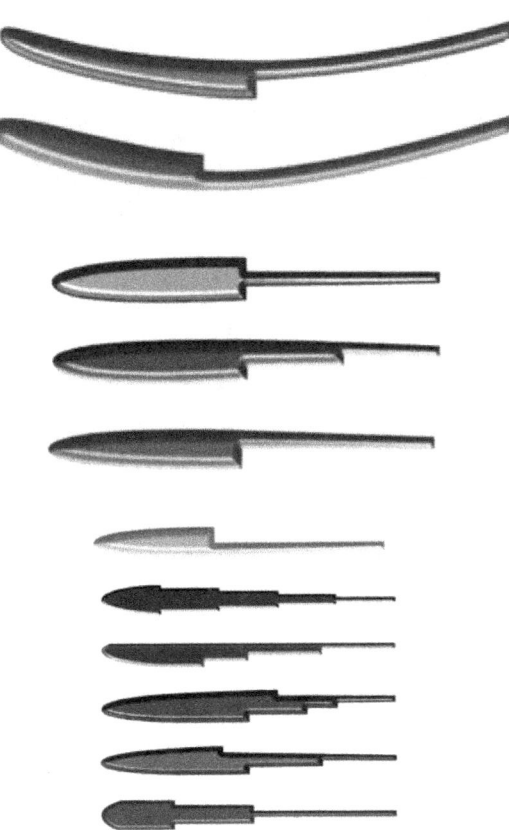

" Coanda principle....".

END PLATES

Now we know the reason for the existence of downforce; it is due to the pressure difference between the top and bottom surfaces, which generate the downward suction.

Due to this pressure difference, and the fundamental laws of nature, the air close to the low pressure area tends to fill this depression trying to equal the pressure above and below the wing; external wing areas are trying to redirect the air from top to bottom; therefore, the wing tip zones are not working as they should:

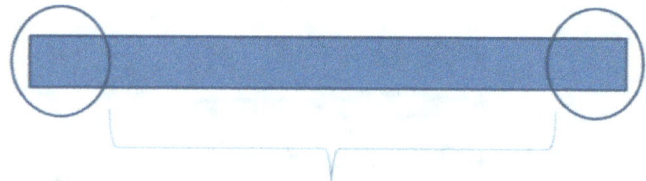

The effective area of suction or downforce generation is reduced; the ideal is that the whole wing generates downforce, for this we put a barrier to the passage of air from top to bottom; this barrier makes the wing work reducing turbulence at tip-ends, reducing the drag:

→ The best:

Transform the profile end wing, in symmetrical; that not produce a pressure difference, so not produce air circulation (vortex in time).

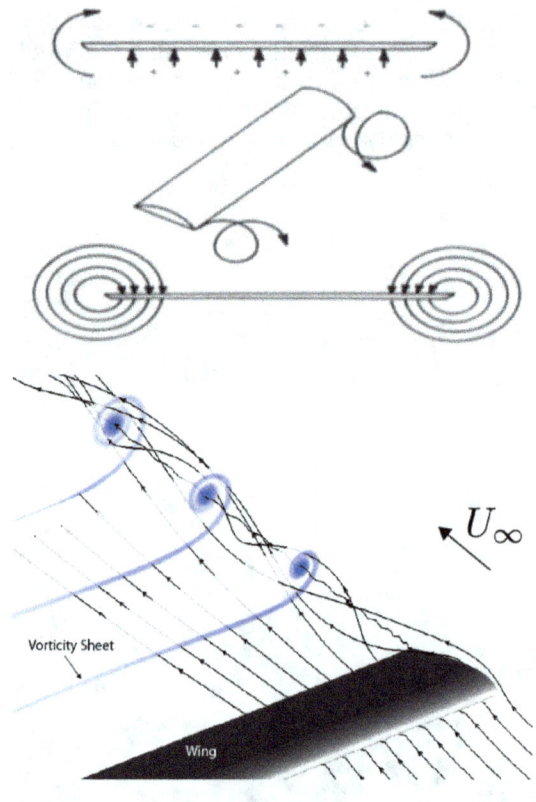

Sure there are many people who do not believe how extreme the turbulence generated by the wings are. To illustrate it we have the following photographs of turbulences generated by a plane (it is normal to think now, that it is dangerous to take off "just" after taking off another plane. ...). This is the reason behind the standard delay between takeoffs, because these vortex can create high turbulences and even structural failure (Wikipedia images):

That is very important in airport protocol; in fact, we can show these graphics:

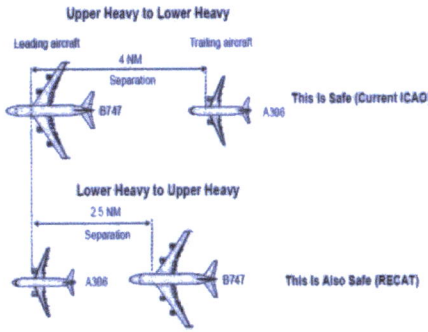

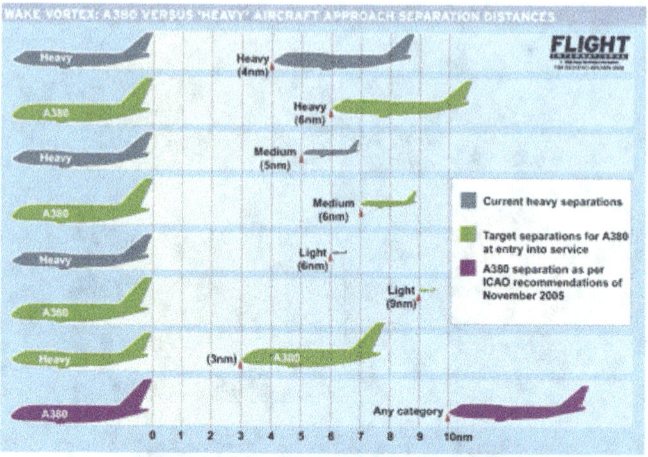

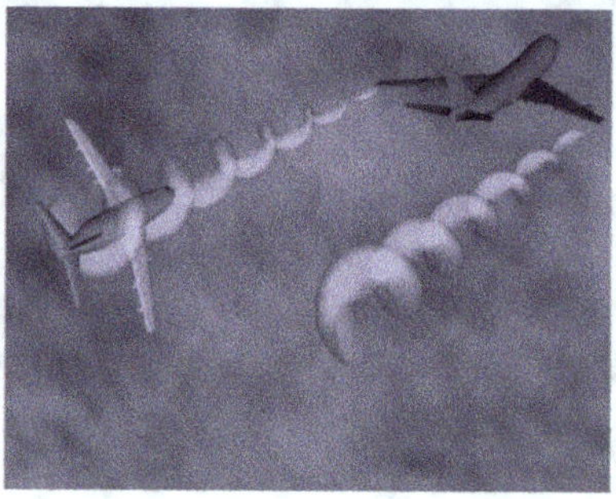

Amazing photo:

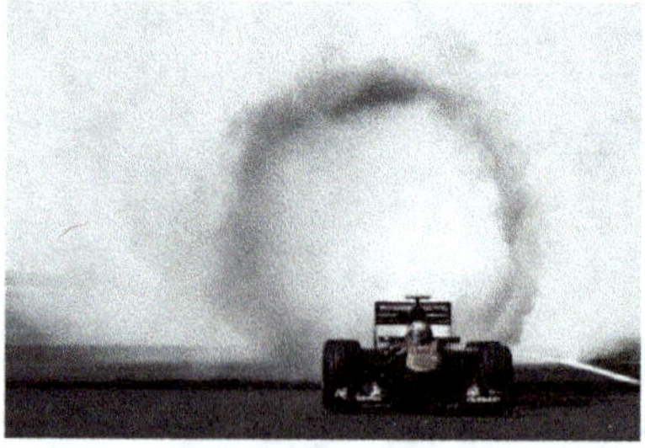

Despite using end-plates, vortices continue to appear, which are smaller and no longer generate drag:

For the design of these end-plates, we can use the "law" that says the height of the endplate should be a certain number of times the thickness of the wing.

With the installation of end plates we can see a virtual increase of the wingspan, let's see how it works:

We know that the aspect ratio, AR = span / chord: scale divided by chord.

We shall base such as a wing with a wingspan of 165 cm and 40 cm chord.

AR = 4.125

There is an expression to calculate the actual or real "AR", once we have installed the end plates:

Effective Aspect Ratio (AR_effective) = Actual Aspect Ratio (AR_actual) * (1 + 1.9 * (endplate depth (mm)/span(cm)))

Applying this expression to our example we get the new "AR" is = 4.837, with a depth of 15 mm endplates. Now, we use this new AR value to calculate the new span, obtaining a value of 193 cm. That is, using endplates, we get that the wingspan is greater (going from 165 cm to 196 cm). Thus, we are generating more downforce, without using larger wings.

Let's see some data taken from track tests to see what happens with the implementation of these end plates.

Suppose that:

Case 1) a wing without end plates.

Case 2) end plates slightly bigger than the wing's thickness.

Case 3) end plates with a height equal to 4 wing's thickness.

Case 4) end plate with a height equal to 5 wing's thickness above the surface of the wing.

We measure downforce, drag and performance (downforce / drag) for each case (force values are in Newtons):

	Downforce	Drag	Lift-to-drag ratio
Case 1	769	195	3.94
Case 2	786	188	4.18
Case 3	873	183	4.77
Case 4	901	178	5.06

We see how the downforce increases, while the drag is reduced.

Here's another study:

Using the previous "generic" wing, we did a CFD study analyzing downforce and drag, with and without end plates; the results shows the following:

$$\frac{CL(with)}{CL(without)} = 1.07$$

$$\frac{CD(with)}{CD(without)} = 0.98$$

As we can see, downforce increases a 7%, while drag is reduced a 2%; values are clearly not "generic" but they do provide an idea of how efficient endplates are in improving performance. In nature, a way of increasing the efficiency of bird wings is also reducing wingtip vortices; for this, birds use a suitable arrangement of feathers at the ends; this causes a reduction of the vortices and thus drag (Wikipedia images):

There are attempts of imitating this by aircrafts:

On the other hand, the following images show a mix between vortex generators and vortex reducers:

We can imitate this device in aircraft wings, paddles, etc:

In any case the objective sought is to reduce marginal vortices and turbulence downstream; this will produce a reduction in drag; in aviation, another system is used for the same purpose; it is the wingtip winglets:

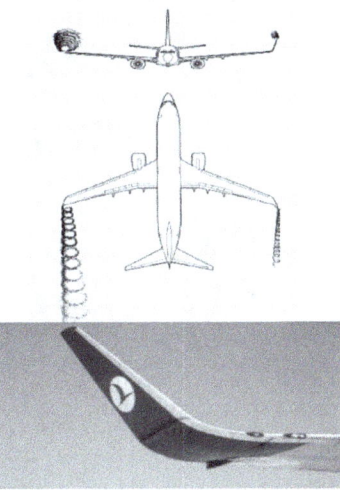

These winglets are sometimes used as a source of aerodynamic thrust, creating a pressure difference between the surfaces of the device, having a component of the resultant force pointing forward of the aircraft.

The principal function of winglets is reducing the drag; that, produce more passengers for example (also fuel reduction).

There are many winglets types, but the goal is reduce the vortex or depression up:

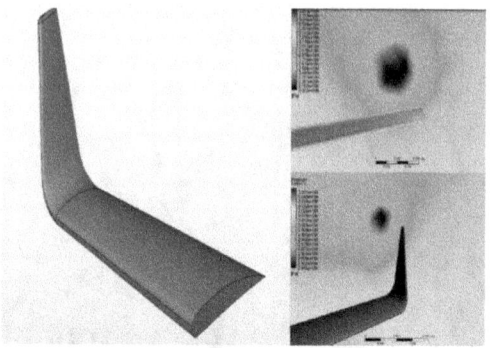

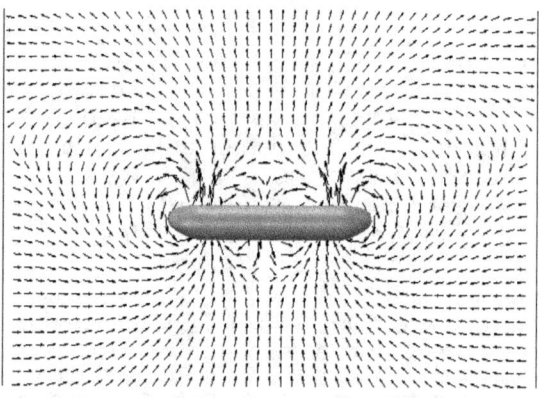

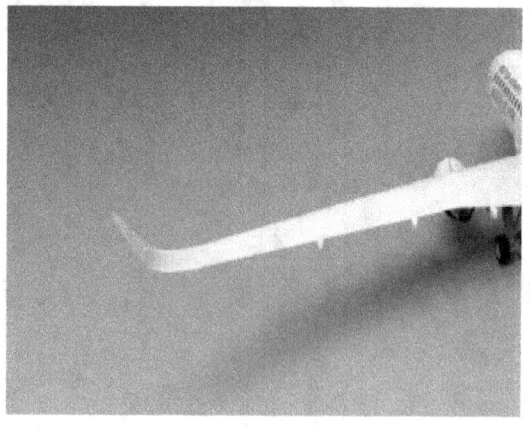

More or less, with winglets devices, is possible to translate 2% passengers more with the same fuel. That is very important. Also, with the same fuel, is possible travel, aprox 4% more kilometers.

➜ We analyzing after, these vortex and their evolution iiii

➜ Vortex amazing in race car:

KÁRMÁN VORTEX STREET

There is a particular type of turbulence or vortices that are created downstream; the reason is different from the turbulence formed at wingtips, but the consequences are the same: increased drag.

The curious thing about these geometries is that we can see them in different contexts and especially in different geometric scales: from contrails of 1cm of diameter up to contrails the size of storms or groups of clouds passing through volcanoes on islands:

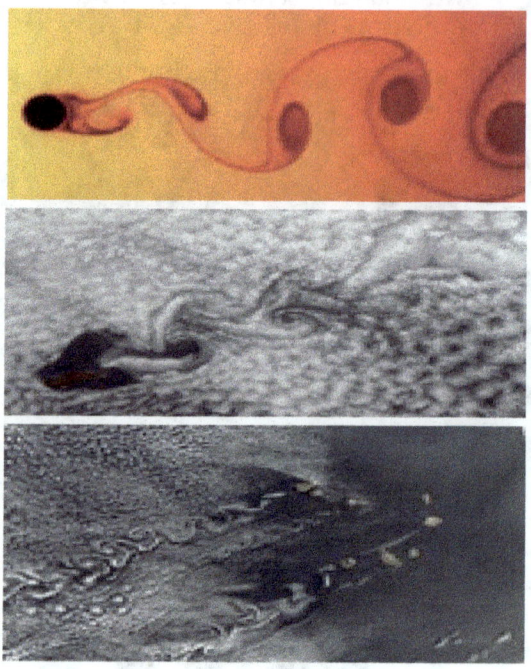

These geometries are formed only under certain conditions; i.e. under a certain Reynolds numbers; we can change the flow rate to obtain them:

These vortices are actually turbulences; the uniqueness lies in its formation, geometry and its geometry being repeated: there is always an equal ratio of width and height of the vortex.

As any turbulence, it should be studied and prevented, producing vibrations that may be detrimental to the performance of the car;

The geometry is the same in all Karman vortices:

$$\frac{b}{a} = \frac{1}{\pi} sinh^{-1}\{1\} = 0.281$$

Because these vortexes have a predictable shape we can use specific instrumentation that reads its velocity. For this type of vortexes there is a direct relationship between fluid velocity and frequency; therefore it is enough to measure the frequency and we'll know speed:

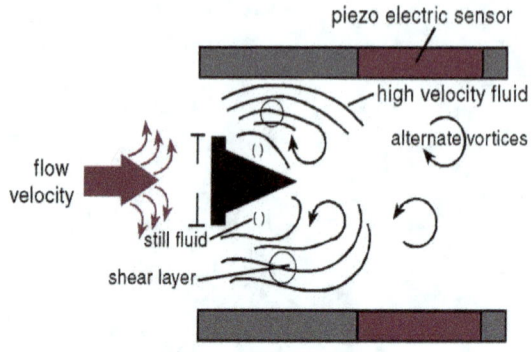

$$S_t = \left(\frac{fd}{V_{avg}} \right)$$

$$Q = AV_{avg} = \frac{fdA}{S_t}$$

where S_t = Strouhal number

f = frequency of the shedding

d = width of the bluff body

V_{avg} = average velocity of the fluid

Q = volumetric flow rate

A = cross-sectional area of the meter body

Abiut the Karman vortex street, there are a lot methods to eliminate the vibrations; that is for example: "D" diameter circle:

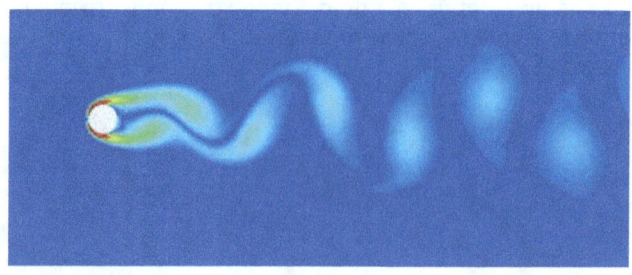

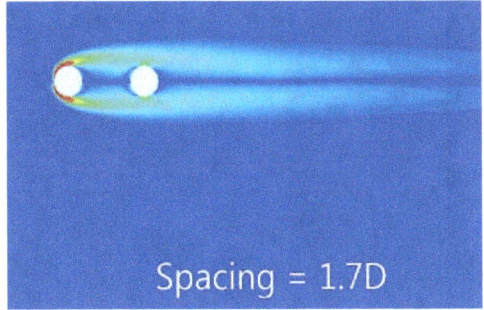

Spacing = 1.7D

USING MULTIPLE WINGS

We know the more wings or flaps we have, more downforce will be generated; this is indeed true, but up to a limit.

By regulation, and assumed there is no restriction on the number of wings that can be used however, the size of the rear wing itself is restricted; therefore it is not possible to physically place a large number of wings. The real limitation should be aerodynamic and not geometric.

See the aerodynamic characteristics of a wing-set, comprising 1, 2, 3 and 4 wing; we calculate their downforce and drag.

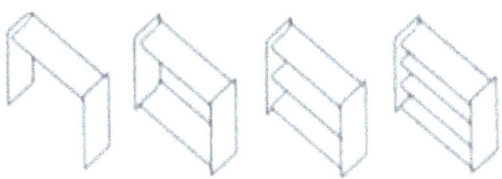

Simon Mc Beath images:

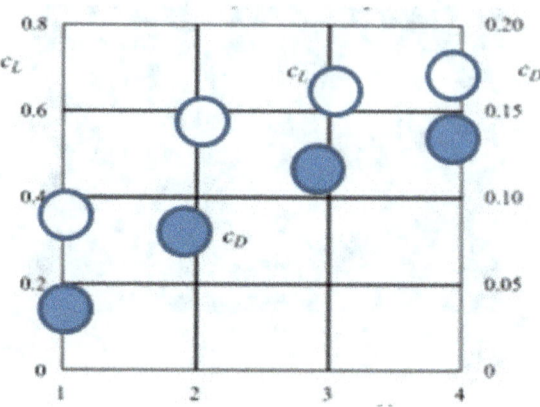

There is a moment where although we increase the number of wings, downforce is not increased, even worse: drag increases dramatically.

In the following graph, we can see that the number of ideal wings is 2.

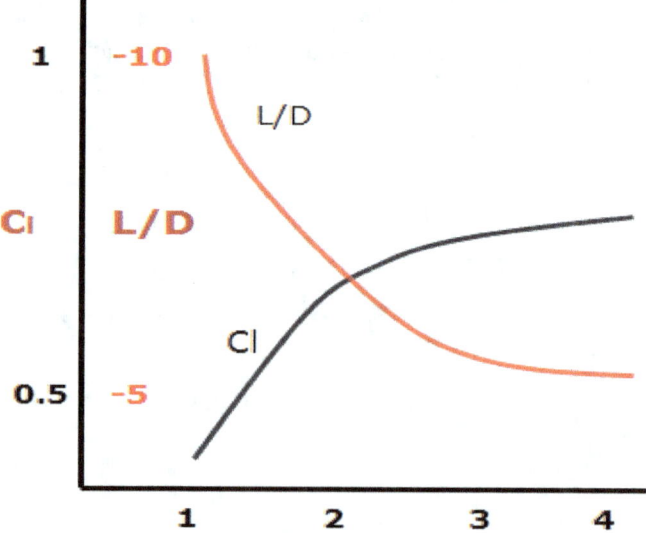

Note:
We know that a F1 car, is symmetrical; seems illogical therefore to put more downforce on one side than the other (except in certain categories where cars are driven in ovals): Williams F1 in the 80s, had more load on the left side; that is, the flap on the front wing on the left side had more impact than the right flap; this was due to the non-symmetry of the aerodynamic loads, and indeed the lap time was reduced. This reduction was also due to the "real" non-symmetry of the car and the circuit characteristics.

SPOILERS

The differentiation between spoiler and wing is simple, not complicated.

A wing is a piece that is responsible for producing downforce while a spoiler makes fewer lift; thus indirectly the spoiler makes downforce, reducing lift.
Given the special shape almost any car has they all generate lift at high speeds (above 200 km/h). This generation of lift can be dangerous when the car is cornering.
The word spoiler, means "breaking"; its function is precisely breaking the circulation of air flow from the top rear of the car, downwards; It is in this way how lift is reduced:

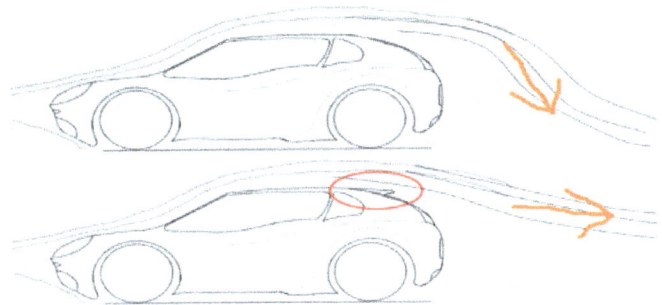

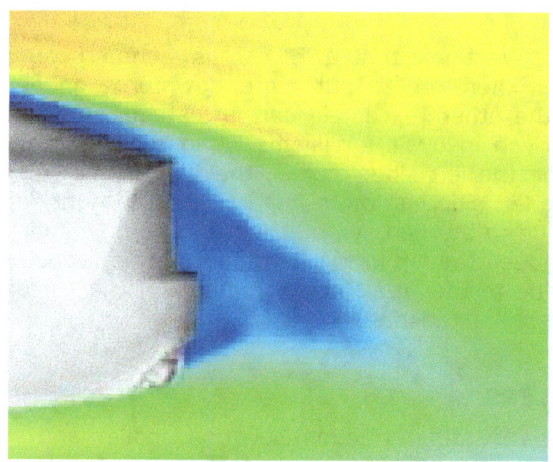

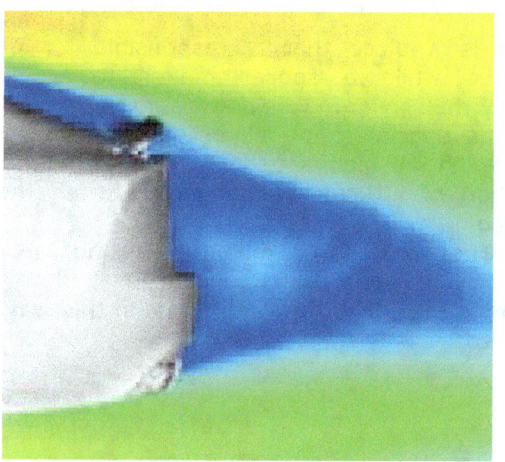

A spoiler has a very important feature that makes it ideal for this role, even more than wings: while wings generate a lot of drag, a spoiler generates very little resistance to the air Simon Mc Beath images:

The truth is that many times, to the common question of what is what? the answer is not easy: what is a wing or a spoiler?

It's a bit complicated, really. In other cases, identification is much easier:

We have all heard about car tuning, and the trend many young people follow tuning cars incorporating aerodynamic features. It's dangerous to do so without the approval of the car's manufacturer and this is very difficult to obtain. It seems perfect to paint the car pink, but incorporating aerodynamic features affects the dynamics of the car and should be done carefully.

Not all cars can place spoilers, in many cases they are used as handles to open the boot but are not strictly spoilers, although they seem it all depends on the final geometry of the car. In the image below, it is useless to install a spoiler in the third car:

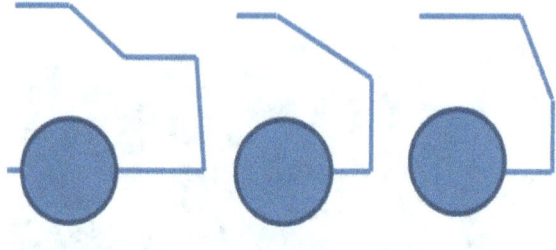

We all know the story of the Audi TT; when it went on sale it had several fatal accidents; studying its aerodynamics carefully, engineers realized that at speeds above 220 km/h the car had a lot of lift; the car tended to take off so the tire grip while cornering was reduced dramatically: Audi had to incorporate a spoiler...

The Audi TT has 2 different types of spoilers; see their downforce in pounds (lbs):
ABT Spoiler:

OEM Spoiler:

90 mph

	FRONT	REAR
ABT	53 lbs	18 lbs
OEM	53 lbs	29 lbs
NO SPOILER	40 lbs	75 lbs

125 mph

	FRONT	REAR
ABT	122 lbs	42 lbs
OEM	122 lbs	67 lbs
NO SPOILER	85 lbs	175 lbs

In any case there is positive lift, but the spoilers reduce the upward force.

The graph below expresses the functionality of a generic spoiler depending on its height at a constant angle (20°) and constant dimensions of the car:

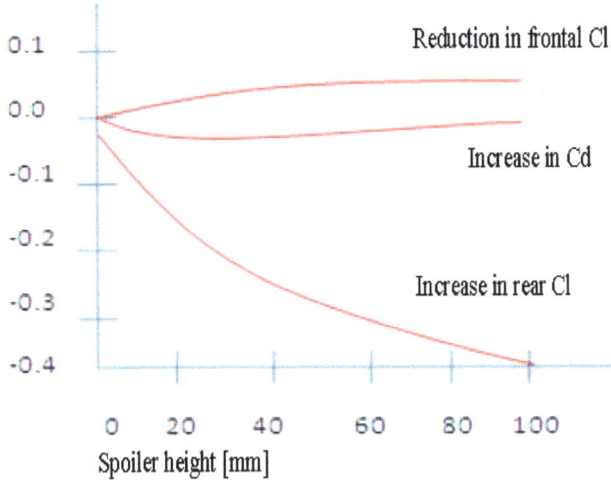

Reduction in frontal Cl

Increase in Cd

Increase in rear Cl

Spoiler height [mm]

We see that drag is almost constant, while downforce on the rear axle increases and the downforce on the front axle are reduced slightly.

Take the case of the Porsche 911; at around 100 km/h, a "wing" that is clearly different from a spoiler rises; There are 911 models in GT categories, which directly have a wing; so the lift created stops being dangerous generating downforce:

WING PROFILES

In the history, the profiles is changed a lot (Wikipedia images):

Evolution of Airfoils

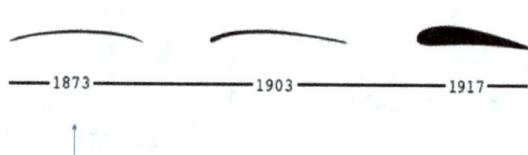

Early Designs - Designers mistakenly believed that these airfoils with sharp leading edges will have low drag.
In practice, they stalled quickly, and generated considerable drag.

We have thought it was proper to include this little section right here to analyze a little bit different profiles, basically within the NACA series; There are very different profiles families with different characteristics: NACA, Youkowski, Eppler, etc ....

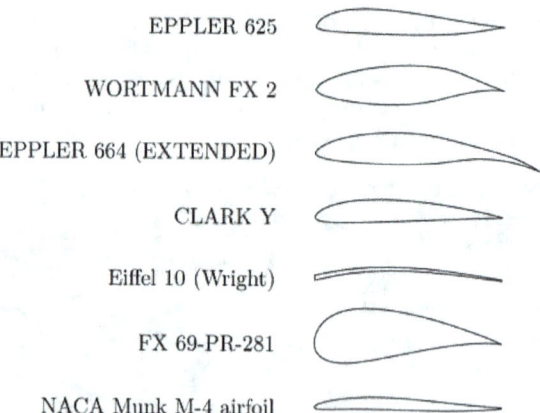

EPPLER 625

WORTMANN FX 2

EPPLER 664 (EXTENDED)

CLARK Y

Eiffel 10 (Wright)

FX 69-PR-281

NACA Munk M-4 airfoil

But perhaps NACA are the easiest to use because they are generated using mathematical equations:

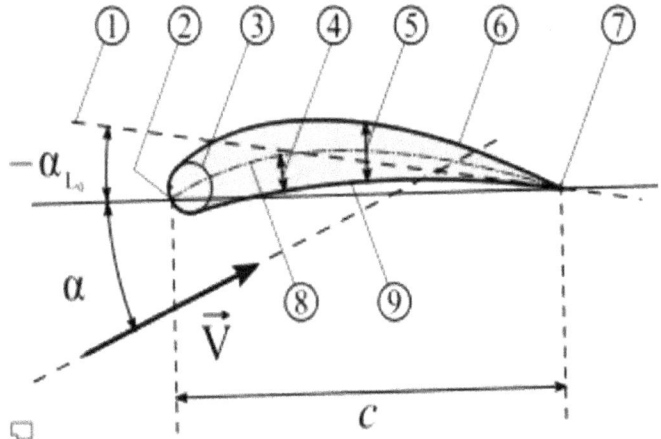

Description: 1: Zero lift line; 2: Leading edge; 3: Nose circle; 4: Camber; 5: Max. thickness; 6: Upper surface; 7: Trailing edge; 8: Camber mean-line; 9: Lower surface.

The following is directly obtained from the http://www.aerospaceweb.org/ site.

The NACA airfoil series

The early NACA airfoil series, the 4-digit, 5-digit, and modified 4-/5-digit, were generated using analytical equations that describe the camber (curvature) of the mean-line (geometric centerline) of the airfoil section as well as the section's thickness distribution along the length of the airfoil. Later families, including the 6-Series, are more complicated shapes derived using theoretical rather tan geometrical methods. Before the National Advisory Committee for Aeronautics (NACA) developed these series, airfoil design was rather arbitrary with nothing to guide the designer except past experience with known shapes and experimentation with modifications to those shapes.

This methodology began to change in the early 1930s with the publishing of a NACA report entitled *The Characteristics of 78 Related Airfoil Sections from Tests in the Variable Density Wind Tunnel.*

In this landmark report, the authors noted that there were many similarities between the airfoils that were most successful, and the two primary variables that affect those shapes are the slope of the airfoil mean camber line and the thickness distribution above and below this line. They then presented a series of equations incorporating these two variables that could be used to generate an entire family of related airfoil shapes. As airfoil design became more sophisticated, this basic approach was modified to include additional variables, but these two basic geometrical values remained at the heart of all NACA airfoil series.

NACA Four-Digit Series:

The first family of airfoils designed using this approach became known as the NACA Four-Digit Series. The first digit specifies the maximum camber (m) in percentage of the chord (airfoil length), the second indicates the position of the maximum camber (p) in tenths of chord, and the last two numbers provide the maximum thickness (t) of the airfoil in percentage of chord.

For example, the NACA 2415 airfoil has a maximum thickness of 15% with a camber of 2% located 40% back from the airfoil leading edge (or 0.4c). Utilizing these m, p, and t values, we can compute the coordinates for an entire airfoil using the following relationships.

1) Pick values of x from 0 to the maximum chord c.
2) Compute the mean camber line coordinates by plugging the values of m and p into the following equations for each of the x coordinates.

$$y_C = \frac{m}{p^2}(2px - x^2) \; from \; x = 0 \; to \; x = p$$

$$y_C = \frac{m}{(1-p)^2}((1-2p) + 2px - x^2) \; from \; x = p \; to \; x = c$$

Where:

x = coordinates along the length of the airfoil, from 0 to c (which stands for chord, or length),

y = coordinates above and below the line extending along the length of the airfoil, these are either yt for thickness coordinates or yc for camber coordinates

t = maximum airfoil thickness in tenths of chord (i.e. a 15% thick airfoil would be 0.15)

m = maximum camber in tenths of the chord.

p = position of the maximum camber along the chord in tenths of chord.

   Calculate the thickness distribution above (+) and below (-) the mean line by plugging the value of t into the following equation for each of the x coordinates.

$$\pm y_t = \frac{t}{0.2}\left(0.2969\sqrt{x} - 0.1260x - 0.3516x^2 + 0.2843x^3 - 0.1015x^4\right)$$

3) Determine the final coordinates for the airfoil upper surface (xU, yU) and lower surface (xL, yL) using the following relationships.

$$x_U = x - y_t \, \sin\theta$$

$$y_U = y_c + y_t \, \cos\theta$$

$$x_L = x + y_t \, \sin\theta$$

$$y_L = y_c - y_t \, \cos\theta$$

$$\text{where } \theta = \arctan\left(\frac{dy_c}{dx}\right)$$

NACA Five-Digit Series:

The NACA Five-Digit Series uses the same thickness forms as the Four-Digit Series but the mean camber line is defined differently and the naming convention is a bit more complex. The first digit, when multiplied by 3/2, yields the design lift coefficient (cl) in tenths. The next two digits, when divided by 2, give the position of the maximum camber (p) in tenths of chord. The final two digits again indicate the maximum thickness (t) in percentage of chord. For example, the NACA 23012 has a maximum thickness of 12%, a design lift coefficient of 0.3, and a maximum camber located 15% back from the leading edge. The steps needed to calculate the coordinates of such an airfoil are:

1) Pick values of x from 0 to the maximum chord c.

2) Compute the mean camber line coordinates for each x location using the following equations, and since we know p, determine the values of m and k1 using the table shown below.

$$y_c = \frac{k_1}{6}(x^3 - 3mx^2 + m^2(3 - m)x) \quad from \; x = 0 \; to \; x = p$$

$$y_c = \frac{k_1 m^3}{6}(1 - x) \quad from \; x = p \; to \; x = c$$

Mean-line designation	Position of max camber (p)	m	k1
210	0.05	0.0580	361.400
220	0.10	0.1260	51.640
230	0.15	0.2025	15.957
240	0.20	0.2900	6.643
250	0.25	0.3910	3.230

3) Calculate the thickness distribution using the same equation as the Four-Digit Series.

4) Determine the final coordinates using the same equations as the Four-Digit Series.

Modified NACA Four- and Five-Digit Series:

The airfoil sections you mention for the B-58 bomber are members of the Four-Digit Series, but the names are slightly different as these shapes have been modified. Let us consider the root section, the NACA 0003.46-64.069, as an example. The basic shape is the 0003, a 3% thick airfoil with 0% camber. This shape is a symmetrical airfoil that is identical above and below the mean camber line.

The first modification we will consider is the 0003-64. The first digit following the dash refers to the roundedness of the nose. A value of 6 indicates that the nose radius is the same as the original airfoil while a value of 0 indicates a sharp leading edge. Increasing this value specifies an increasingly more rounded nose. The second digit determines the location of maximum thickness in tenths of chord. The default location for all four- and five-digit airfoils is 30% back from the leading edge. In this example, the location of maximum thickness has been moved back to 40% chord.

Finally, notice that the 0003.46-64.069 features two sets of digits preceded by decimals. These merely indicate slight adjustments to the maximum thickness and location thereof. Instead of being 3% thick, this airfoil is 3.46% thick. Instead of the maximum thickness being located at 40% chord, the position on this airfoil is at 40.69% chord. To compute the coordinates for a modified airfoil shape:

1) Pick values of x from 0 to the maximum chord c.
2) Compute the mean camber line coordinates using the same equations provided for the Four or Five-Digit Series as appropriate.
3) Calculate the thickness distribution above (+) and below (-) the mean line using these equations. The values of the ax and dx coefficients are determined from the following table (these are derived for a 20% thick airfoil).

$$y_t = a_0\sqrt{x} + a_1 x + a_2 x^2 + a_3 x^3 \text{ ahead of } t_{max}$$

$$y_t = d_0 + d_1(1-x) + d_2(1-x)^2 + d_3(1-)x^3 \text{ aft of } t_{max}$$

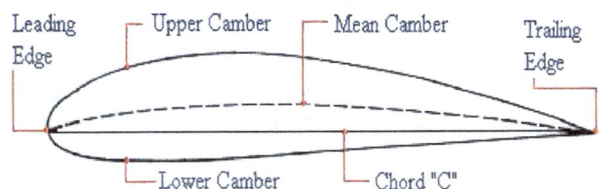

Airfoil	a_0	a_1	a_2	a_3	d_0	d_1	d_2	d_3
0020-62	0.296900	0.213337	-2.931954	5.229170	0.002000	0.200000	-0.040625	-0.070312
0020-63	0.296900	-0.096082	-0.543310	0.559395	0.002000	0.234000	-0.068571	-0.093878
0020-64	0.296900	-0.246867	0.175384	-0.266917	0.002000	0.315000	-0.233333	-0.032407
0020-65	0.296900	-0.310275	0.341700	-0.321820	0.002000	0.465000	-0.684000	0.292000
0020-66	0.296900	-0.271180	0.140200	-0.082137	0.002000	0.700000	-1.662500	1.312500
0020-03	0.000000	0.920286	-2.801900	2.817990	0.002000	0.234000	-0.068571	-0.093878
0020-33	0.148450	0.412103	-1.672610	1.688690	0.002000	0.234000	-0.068571	-0.093878
0020-93	0.514246	-0.840115	1.110100	-1.094010	0.002000	0.234000	-0.068571	-0.093878
0020-05	0.000000	0.477000	-0.708000	0.308000	0.002000	0.465000	-0.684000	0.292000
0020-35	0.148450	0.083362	-0.183150	-0.006910	0.002000	0.465000	-0.684000	0.292000
0020-34	0.148450	0.193233	-0.558166	0.283208	0.002000	0.315000	-0.233333	-0.032407

4) Determine the "final" coordinates using the same equations as the Four-Digit Series.
5) As noted above, this procedure yields a 20% thick airfoil. To obtain the desired thickness, simply scale the airfoil by multiplying the "final" y coordinates by **[t / 0.2]**.

NACA 1-Series or 16-Series:

Unlike those airfoil families discussed so far, the 1-Series was developed based on airfoil theory rather than on geometrical relationships. By the time these airfoils were designed during the late 1930s, many advances had been made in inverse airfoil design methods. The basic concept behind this design approach is to specify the desired pressure distribution over the airfoil (this distribution dictates the lift characteristics of the shape) and then derive the geometrical shape that produces this pressure distribution. As a result, these airfoils were not generated using some set of analytical expressions like the Four- or Five-Digit Series.

The 1-Series airfoils are identified by five digits, as exemplified by the NACA 16-212. The first digit, 1, indicates the series (this series was designed for airfoils with regions of barely supersonic flow).

The 6 specifies the location of minimum pressure in tenths of chord, i.e. 60% back from the leading edge in this case. Following a dash, the first digit indicates the design lift coefficient in tenths (0.2) and the final two digits specify the maximum thickness in tenths of chord (12%).

Since the 16-XXX airfoils are the only ones that have ever seen much use, this family is often referred to as the 16-Series rather than as a subset of the 1-Series.

NACA 6-Series:

Although NACA experimented with approximate theoretical methods that produced the 2-Series through the 5-Series, none of these approaches was found to accurately produce the desired airfoil behavior. The 6-Series was derived using an improved theoretical method that, like the 1-Series, relied on specifying the desired pressure distribution and employed advanced mathematics to derive the required geometrical shape. The goal of this approach was to design airfoils that maximized the region over which the airflow remains laminar. In so doing, the drag over a small range of lift coefficients can be substantially reduced. The naming convention of the 6-Series is by far the most confusing of any of the families discussed thus far, especially since many different variations exist. One of the more common examples is the NACA 641-212, a=0.6.

In this example, 6 denotes the series and indicates that this family is designed for greater laminar flow than the Four- or Five-Digit Series. The second digit, 4, is the location of the minimum pressure in tenths of chord (0.4c). The subscript 1 indicates that low drag is maintained at lift coefficients 0.1 above and below the design lift coefficient (0.2) specified by the first digit after the dash in tenths.

The final two digits specify the thickness in percentage of chord, 12%. The fraction specified by a=___ indicates the percentage of the airfoil chord over which the pressure distribution on the airfoil is uniform, 60% chord in this case. If not specified, the quantity is assumed to be 1, or the distribution is constant over the entyre airfoil.

NACA 7-Series:

The 7-Series was a further attempt to maximize the regions of laminar flow over an airfoil differentiating the locations of the minimum pressure on the upper and lower surfaces. An example is the NACA 747A315. The 7 denotes the series, the 4 provides the location of the minimum pressure on the upper surface in tenths of chord (40%), and the 7 provides the location of the minimum pressure on the lower surface in tenths of chord (70%). The fourth character, a letter, indicates the thickness distribution and mean line forms used. A series of standardized forms derived from earlier families are designated by different letters. Again, the fifth digit incidates the design lift coefficient in tenths (0.3) and the final two integers are the airfoil thickness in percentage of chord (15%).

NACA 8-Series:

A final variation on the 6- and 7-Series methodology was the NACA 8-Series designed for flight at supercritical speeds. Like the earlier airfoils, the goal was to maximize the extent of laminar flow on the upper and lower surfaces independently. The naming convention is very similar to the 7-Series, an example being the NACA 835A216. The 8 designates the series, 3 is the location of minimum pressure on the upper surface in tenths of chord (0.3c), 5 is the location of minimum pressure on the lower surface in tenths of chord (50%), the letter A distinguishes airfoils having different camber or thickness forms, 2 denotes the design lift coefficient in tenths (0.2), and 16 provides the airfoil thickness in percentage of chord (16%).

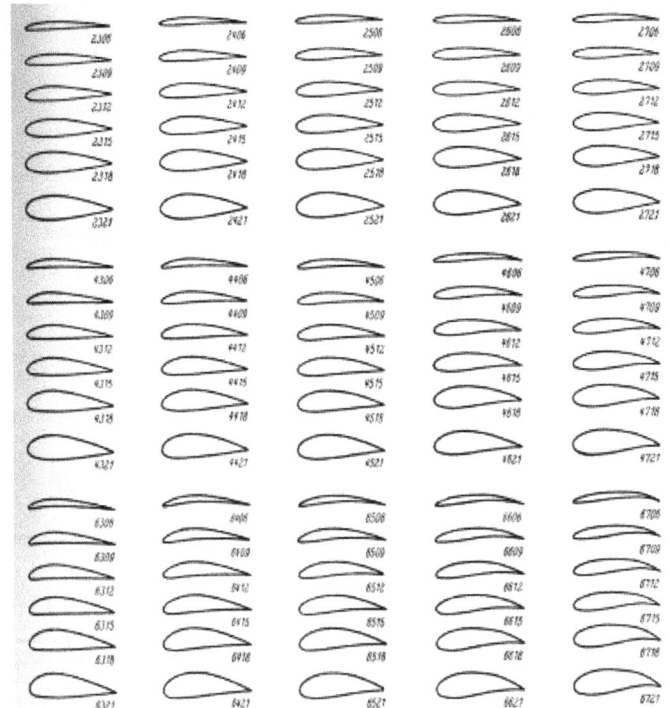

Summary:

Though we have introduced the primary airfoil families developed in the United States before the advent of supersonic flight, we haven't said anything about their uses. So let's briefly explore the advantages, disadvantages, and applications of each of these families.

Family	Advantages	Disadvantages	Applications
4-Digit	1. Good stall characteristics 2. Small center of pressure movement across large speed range 3. Roughness has little effect	1. Low maximum lift coefficient 2. Relatively high drag 3. High pitching moment	1. General aviation 2. Horizontal tails Symmetrical: 3. Supersonic jets 4. Helicopter blades 5. Shrouds 6. Missile/rocket fins
5-Digit	1. Higher maximum lift coefficient 2. Low pitching moment 3. Roughness has little effect	1. Poor stall behavior 2. Relatively high drag	1. General aviation 2. Piston-powered bombers, transports 3. Commuters 4. Business jets
16-Series	1. Avoids low pressure peaks 2. Low drag at high speed	1. Relatively low lift	1. Aircraft propellers 2. Ship propellers
6-Series	1. High maximum lift coefficient 2. Very low drag over a small range of operating conditions 3. Optimized for high speed	1. High drag outside of the optimum range of operating conditions 2. High pitching moment 3. Poor stall behavior 4. Very susceptible to roughness	1. Piston-powered fighters 2. Business jets 3. Jet trainers 4. Supersonic jets
7-Series	1. Very low drag over a small range of operating conditions 2. Low pitching moment	1. Reduced maximum lift coefficient 2. High drag outside of the optimum range of operating conditions 3. Poor stall behavior 4. Very susceptible to roughness	Seldom used
8-Series	Unknown	Unknown	Very seldom used

Today, airfoil design has in many ways returned to an earlier time before the NACA families were created. The computational resources available now allow the designer to quickly design and optimize an airfoil specifically tailored to a particular application rather than making a selection from an existing family.

High-lift devices	Increase of maximum lift	Angle of basic aerofoil at max. lift	Remarks
Double-slotted Fowler flap	100%	20°	Same as Fowler flap only more so. Treble slots sometimes used.
Krueger flap	50%	25°	Nose-flap hinging about leading edge. Reduces lift at small deflections. Nose-up pitching moment.
Slotted wing	40%	20°	Controls boundary layer. Slight extra drag at high speeds.
Fixed slat	50%	20°	Controls boundary layer. Extra drag at high speeds. Nose-up pitching moment.

Movable slat	60%	22°	Controls boundary layer. Increases camber and area. Greater angles of attack. Nose-up pitching moment.
Slat and slotted flap	75%	25°	More control of boundary layer. Increased camber and area. Pitching moment can be neutralized.
Slat and double-slotted Fowler flap	120%	28°	Complicated mechanisms. The best combination for lift; treble slots may be used. Pitching moment can be neutralized.
Blown flap	80%	16°	Effect depends very much on details of arrangement.
Jet flap	60%	?	Depends even more on angle and velocity of jet.

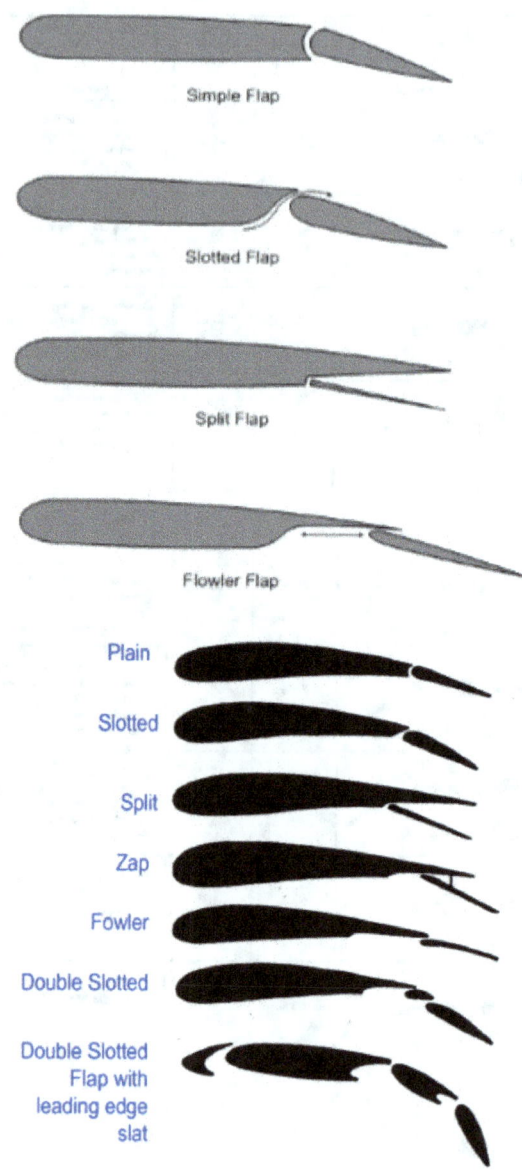

Simple Flap

Slotted Flap

Split Flap

Flowler Flap

Plain

Slotted

Split

Zap

Fowler

Double Slotted

Double Slotted
Flap with
leading edge
slat

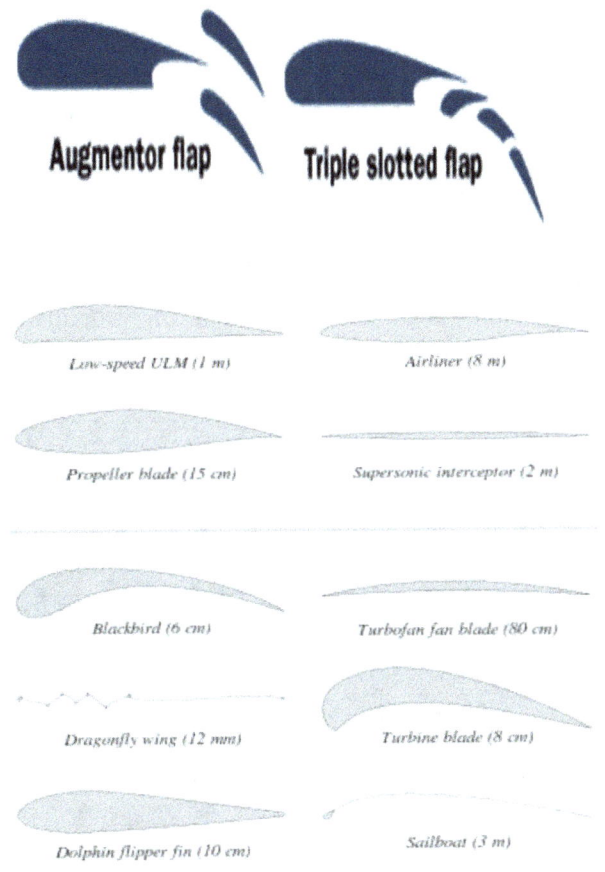

Augmentor flap

Triple slotted flap

Low-speed ULM (1 m)

Airliner (8 m)

Propeller blade (15 cm)

Supersonic interceptor (2 m)

Blackbird (6 cm)

Turbofan fan blade (80 cm)

Dragonfly wing (12 mm)

Turbine blade (8 cm)

Dolphin flipper fin (10 cm)

Sailboat (3 m)

Other profiles like Youkowsky (will be discussed below) that have a mathematical facility to scan them, but their effectiveness is reduced.

CALCULATING DOWNFORCE FROM TRACK DATA

Let's do a "real" test calculating a car's downforce from track data.

The displacement of the dampers or springs, are due to:

- Longitudinal, transverse and vertical accelerations.
- Fuel consumption.

Therefore, if the displacement or elongation of the dampers subtracts those due to these two factors, we are left with the displacement generated by downforce. Knowing the suspension parameters, we can calculate the loads on each wheel.

From the elongations of the dampers, we can know the load on the car, knowing the parameters of the springs; if we know that each "mm" elongation corresponds to "x" kilos, it's easy: Estoril circuit:

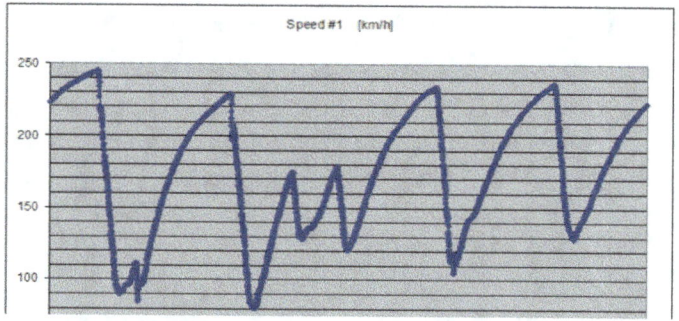

What we do now is to choose and display a line stretching:

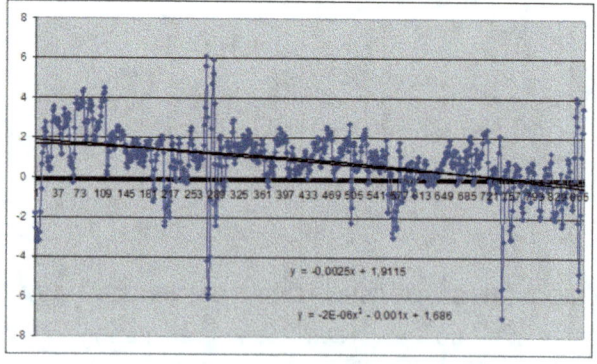

We see that interpolating these values tends to decrease (to compress the damper.)

26.78 kg / mm on FRONT

30.35 kg / mm on REAR

V	mm	mm	
135.0	24.4	16.3	
210.0	16.1	8.6	
	8.3	7.7	← **Difference**

600 kg at 210 km/h.

We can improve this process, to get much more accuracy; let's subtract the signal dampers elongation, elongations caused by accelerations (3D) and fuel consumption:

Normally, we get graphs of the data that looks like chaos...

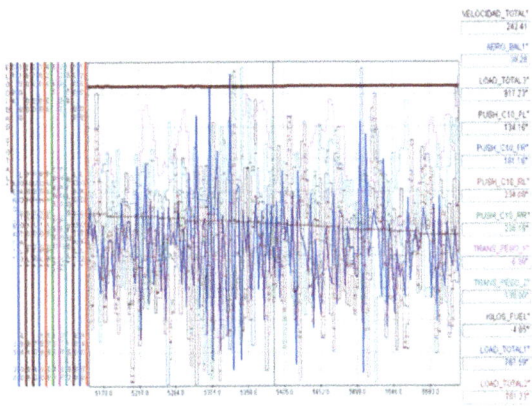

You have to get only the following data: speed, balance and total load:

Mathematicians channels are:

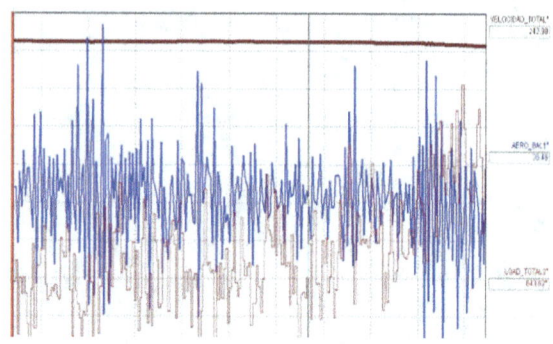

1) AERO_BAL1: Freq= 20Hz Ndec=2 Dec

$AERO\_10 = ((PUSH\_C10\_FL + WEIGHT\_NOPUSH\_FL + PUSH\_C10\_FR + WEIGHT\_NOPUSH\_FR)/(PUSH\_C10\_RL + WEIGHT\_NOPUSH\_RL + PUSH\_C10\_RR + WEIGHT\_NOPUSH\_RR + PUSH\_C10\_FL + WEIGHT\_NOPUSH\_FL + PUSH\_C10\_FR + WEIGHT\_NOPUSH\_FR))*100$

2) DF_FRONT: Freq= 20Hz Ndec=2 Dec

$DF\_FRONT = PUSH\_C10\_FL + WEIGHT\_NOPUSH\_FL + PUSH\_C10\_FR + WEIGHT\_NOPUSH\_FR$

3) DF_REAR: Freq= 20Hz Ndec=2 Dec

$DF\_REAR = PUSH\_C10\_RL + WEIGHT\_NOPUSH\_RL + PUSH\_C10\_RR + WEIGHT\_NOPUSH\_RR$

4) KILOS_FUEL: Freq= 20Hz Ndec=2 Dec

$$KILOS\_FUEL = (FUEL + REF\_01 +$$
$$REF\_02 + REF\_03 + REF\_04 +$$
$$REF\_05 + REF\_06 + REF\_07 +$$
$$REF\_08 + REF\_09 + REF\_10 +$$
$$F\_TACL)$$

⇒ These are the weights of the different tests and refueling in others.

    5) LOAD_TOTAL1: Freq= 20Hz Ndec=2 Dec

$$LOAD\_TOTAL1=PUSH\_C10\_FL+WE$$
$$IGHT\_NOPUSH\_FL+PUSH\_C10\_FR$$
$$+$$
$$WEIGHT\_NOPUSH\_FR+PUSH\_C10\_$$
$$RL+WEIGHT\_NOPUSH\_RL+PUSH\_C$$
$$10\_RR+ WEIGHT\_NOPUSH\_RR$$

    6) PUSH_C10_FL: Freq= 20Hz Ndec=2 Dec

$$PUSH\_C10\_FL = (MD0\_13 *$$
$$PRFL\_A) + PRFL\_B$$

    7) PUSH_C10_FR: Freq= 20Hz Ndec=2 Dec

$$PUSH\_C10\_FR = (MD0\_14 *$$
$$PRFR\_A) + PRFR\_B$$

    8) PUSH_C10_RL: Freq= 20Hz Ndec=2 Dec

$$PUSH\_C10\_RL = (MD1\_12 *$$
$$PRRL\_A) + PRRL\_B$$

    9) PUSH_C10_RR: Freq= 20Hz Ndec=2 Dec

$$PUSH\_C10\_RR = (MD1\_13 *$$
$$PRRR\_A) + PRRR\_B$$

⇒ Weight transfer in the "X" axis:

10) TRANS_ WEIGHT _X: Freq=   20Hz Ndec=2
   Dec

$$TRANS\_ WEIGHT\_X = (ACC\_X * (CW\_FL + CW\_RL + CW\_FR + CW\_RR) * H) / 2980$$

⇒ 2980 mm, corresponding to the wheelbase of the car in question ...

⇒ Transfers weight in the axis "Z":

11) TRANS_ WEIGHT _Z_FL: Freq=   20Hz Ndec=2
   Dec

$$TRANS\_ WEIGHT\_Z\_FL = ACC\_Z * CW\_FL$$

12) TRANS_ WEIGHT _Z_FR: Freq=   20Hz Ndec=2
   Dec

$$TRANS\_ WEIGHT\_Z\_FR = ACC\_Z * CW\_FR$$

13) TRANS_ WEIGHT _Z_RL: Freq=   20Hz Ndec=2
   Dec

$$TRANS\_ WEIGHT\_Z\_RL = ACC\_Z * CW\_RL$$

14) TRANS_ WEIGHT _Z_RR: Freq=   20Hz Ndec=2
   Dec

$$TRANS\_ WEIGHT\_Z\_RR = ACC\_Z * CW\_RR$$

⇒ Total average speed of the 4 wheel speeds:

15) TOTAL_SPEED: Freq= 20Hz Ndec=2 Dec

$$TOTAL\_SPEED = (FL\_WSP + FR\_WSP + RL\_WSP + RR\_WSP) / 4$$

16) WEIGHT_NOPUSH_FL: Freq= 20Hz Ndec=0 Dec

$$WEIGHT\_NOPUSH\_FL = -CW\_FL+(TRANS\_WEIGHT\_X/2)+TRANS\_WEIGHT\_Z\_FL$$

17) WEIGHT_NOPUSH_RL: Freq= 20Hz Ndec=0 Dec

$$WEIGHT\_NOPUSH\_RL = -CW\_RL - (TRANS\_WEIGHT\_X/2) + TRANS\_WEIGHT\_Z\_RL$$

18) WEIGHT_NOPUSH_RR: Freq= 20Hz Ndec=0 Dec

$$WEIGHT\_NOPUSH\_RR = -CW\_RR - (TRANS\_WEIGHT\_X/2) + TRANS\_WEIGHT\_Z\_RR$$

Sensor calibration, for the 2 parameters that make up the line "calibration" of each sensor elongation, is necessary; performing a test load (kg) for a sensor signal (in this case, millivolts):

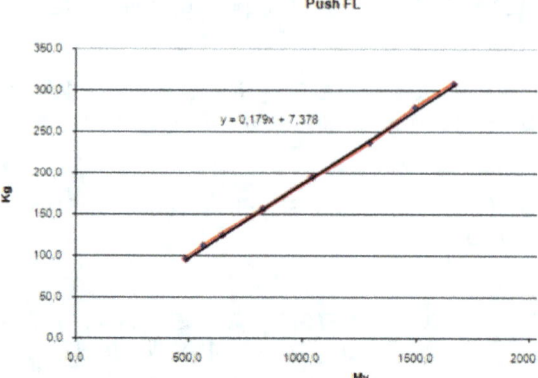

Not all lines of sensor calibration are equal, or even straight...:

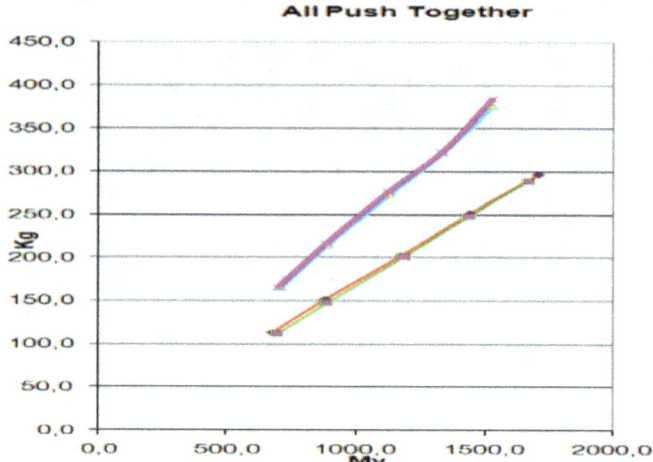

DEFLECTION OF A WING, KNOWING DOWNFORCE

Let's see how to calculate the deflection of a wing, assuming a tip load:

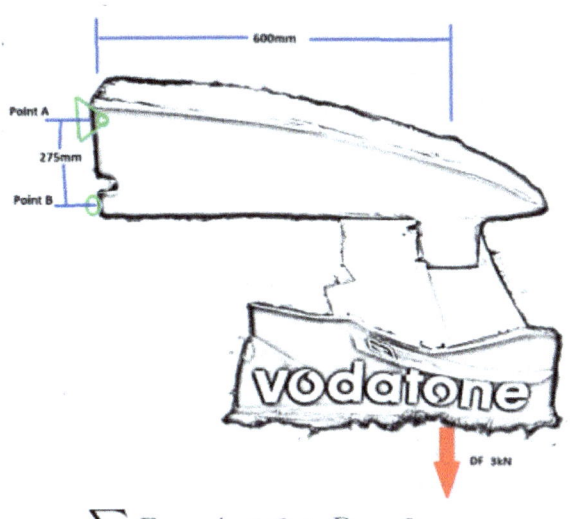

$$\sum F_x = A_x + 0 + B_x = 0$$

$$\sum F_y = 0 + A_y + 0 - 3000 = 0$$

$$\sum M_A = .275B_x - 1800 = 0$$

$$\begin{bmatrix} 1 & 0 & 1 \\ 0 & 1 & 0 \\ 0 & 0 & .275 \end{bmatrix} * \begin{bmatrix} A_x \\ A_y \\ B_x \end{bmatrix} = \begin{bmatrix} 0 \\ 3000 \\ 1800 \end{bmatrix}$$

$$A_x = -6.55kN$$
$$A_y = 3.00kN$$
$$B_x = 6.55kN$$

LAP TIME – DOWNFORCE

Knowing the required downforce for a given car in a particular circuit is a necessary task, but a complicated one; For this purpose, lap-time simulators are used; they are not more than specific programs that calculate lap times depending on "all" the parameters that define a car.

We have decided to include here the design of a simple Lap Time because with this tool we can see the influence of the values of downforce on the dynamics of a car. This first Lap Time simulator has no account of the difference in height of circuit sectors, we also assume that the load is constant. Lap Time simulators that simulate a planar circuit provide great advantages for its speed, simplicity and accuracy.

If we wanted to simulate height variations of a circuit, we need to change the available engine power depending on the slope. These are the expressions for a 3D Lap time simulator:

ENTER DATA:
A = 1.36 Section area.
p = 1.23 Air Density.
Cx = 0.53 Aerodynamics Drag Coefficient.
m = 5/100 Slope Land.
> **Vv:** = 15/36 component of the wind speed in the direction of displacement. km/h, is positive when it is contrary to the movement direction.
M = 1330 mass of the vehicle.

$$V = 0{,}28/3.6,....,280/3.6 \rightarrow \text{Velocity (km/h)}$$

$$Pa(V) = \frac{1}{2}\frac{\rho}{9.8} A \frac{Cx}{75}\left(V+Vv\right)^3 \rightarrow \text{Aerodynamic Power}$$

$$Pr(V) = M\left[(0.012+0.003\left(V+Vv\right)^{11})\frac{V+Vv}{75}\right] \rightarrow \text{Sliding Power}$$

$$Pp(V) = Mm\frac{V+Vv}{75} \rightarrow \text{Slope Power}$$

$$Pt(V) = Pa(V)+Pr(V)+Pp(V)$$

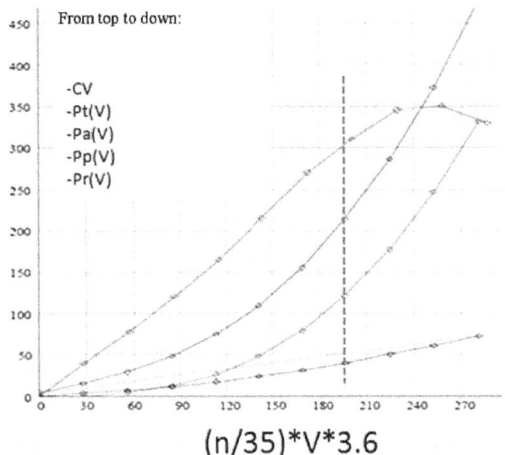

From top to down:

-CV
-Pt(V)
-Pa(V)
-Pp(V)
-Pr(V)

$$(n/35)*V*3.6$$

To have a 3D Lap Time simulator, we can choose an alternative procedure from the one explained above: We modify the power curves as follows:

1) Transforming torque to driving force. We need the power curve of the engine for each gear (knowing the gearing) and taking into account the dynamic tire radius (depending on speed), so that a table of driving forces will be taken depending on the speed vehicle.

2) In contrast to the driving force direction (subtracting), we have drag, we know, aerodynamic, mechanical losses of the entire driveline (i.e. not whether the power curve comes from dynamometer and has not been corrected), and tire friction losses, but now having slopes there is an added force that equals the weight multiplied by the sine of the angle of the slope (if the slope is downward is negative and the operating result is a force for progress).

3) The difference between the driving force under the drag force is the force which accelerates the car (2nd Law of Newton F = m * a); When two forces are equal the car no longer

accelerates.

Let's see how to calculate the extra power of a car, when going up a slope knowing the speed variation:

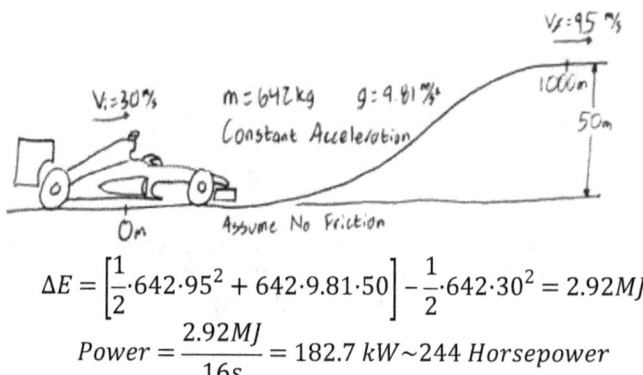

$$\Delta E = \left[\frac{1}{2}\cdot 642\cdot 95^2 + 642\cdot 9.81\cdot 50\right] - \frac{1}{2}\cdot 642\cdot 30^2 = 2.92MJ$$

$$Power = \frac{2.92MJ}{16s} = 182.7\ kW \sim 244\ Horsepower$$

If you need to simulate a car on rainy days, we should vary the friction coefficient and reduce engine power approximately 15%, depending on the amount of water...

Lap time simulation is a useful tool to understand the impact that every factor has on lap time in order to define a starting point for the setup of the vehicle and know which parameters need to be taken care of to define strategies.

The aim of this exercise is for the student to understand how simulation can assist in making decisions by analyzing the importance of each of the following factors on lap time, the factors are: Vehicle mass; Engine power; CG height; Track width; Downforce and Aero efficiency (L/D).

Build a spreadsheet in Excel and vary each variable + and – 10%, record the change in lap time and present the results as clear as possible. State any conclusions from the results.

Circuit:

The circuit is comprised of three straights of 500m each intercalated with three corners all with different radius.

Variables:

Vehicle Mass, m = 660 kg
Engine Power, P = 600 kW
CG Height, h = 0,25 m
Track Width, track = 1,6 m
Downforce = 10,5 kN
Aero Efficiency, aero_eff = 3
Drag Reference, drag_ref =
downforce/aero_eff   (units in kN)
Reference Velocity, V_ref = 70 m/s
Corner radius, R:
Corner 1 = 120 m
Corner 2 = 180 m
Corner 3 = 240 m

Assumption:
- All the static weight is equally distributed over all the wheels.
- The downforce is also equally distributed over all the wheels.
- For the corner distance assume that each corner is 1/3 of the full circle.
- Reference velocity is the velocity at which the downforce is quoted.

Because aerodynamic forces change proportionally to the square of the velocity, given the velocity (V_ref) allows us to calculate the aerodynamic force at any velocity.

Aero efficiency is the ratio between downforce and drag.

CG Height is the height of the center of gravity of the vehicle.

Cornering:

The cornering speed is either limited by the grip that the tires can generate (traction limited) or by the power produced by the engine (power limited).

Power Limited:

The engine cannot produce enough power to overcome the drag at the maximum cornering speed allowed by the grip of the tires.

$$Available\_Power = \frac{P \times 1000 - D_{total} \times V}{1000}$$

[kW] ....(1)

Where:

$$D_{total} = D_{tyre} + D_{aero}$$

[N] .... (2)

D_{total}, total drag force
D_{tyre}, drag from the tires
D_{aero}, aerodynamic drag
V, velocity

$$D_{aero} = \frac{V^2}{V\_ref^2} \times drag\_ref \times 1000$$

[N]

.... (3)

$$D_{tyre} = m \times \frac{v^2}{R} \times \tan\alpha + C_r \times m \times g$$

[N] .... (4)

Where Cr is the rolling coefficient (use $Cr = 0.02$) and α is the slip angle at which the tire generates maximum lateral force (Assume α= 6º).

Traction limited:

$$F_{y\_max} = F_{y\_required}$$

[N] .... (5)

$F_{y\_required}$ is the force that the tires have to match for the system to attain equilibrium, in this case the system is the vehicle cornering at the limit; this force comes from lateral acceleration.

$$F_{y\_required} = m \times a_y = m \times \frac{V^2}{R}$$

[N] .... (6)

F_{y_max} is the maximum lateral force that all four tires can generate.

During cornering, the tires on the side of the car closer to the center of the corner are called inside tires, and the tires on the side of the car farther to the center of the corner are called outside tires:

$$F_{y\_max} = F_{yo} \times 2 + F_{yi} \times 2$$

[N] .... (7)

F_{yo} is the lateral force generated by each of the outside tires.

F_{yi} is the lateral force generated by each of the inside tires.

The maximum lateral force generated from a tire can be derived from the vertical force at the *contact patch*, which is the contact surface between the tire and the ground.

$$F_y = F_z \times \mu$$

[N] ....

(8)

$$\mu = 1.82 - \frac{\left(\dfrac{F_z}{1000}\right)^2}{45}$$

.... (9)

The equation to obtain μ for this case is an empirical equation obtained experimentally.

Although this equation is only valid for a specific tire it implies that μ varies with load which is applicable for all tires.

The vertical force acting on the ground at the contact patch is the sum of aerodynamic, static and dynamic forces (load transfer).

$$F_z = F_{z\_aero} + F_{z\_static} + F_{z\_load\_transfer}$$

[N] .... (10)

The difference between the vertical loads on the inside and outside tires is the load transfer during cornering. For F_{zo} load transfer is positive and for F_{zi} load transfer is negative.

$$F_{Z\_Load\_transfer} = \frac{m \times h \times V^2}{2 \times track \times R}$$ [N]

.... (11)

Knowing that the static weight is equally distributed over the 4 wheels, static load at each wheel is:

$$F_{z\_static} = \frac{m \times g}{4}$$

[N] .... (12)

Similarly to Aerodynamic drag (equation (3)), the Aerodynamic load is:

$$F_{z\_aero} = \frac{\frac{V^2}{V\_ref^2} \times downforce \times 1000}{4}$$ [N] .... (13)

With the difference that the reference force used is the *downforce* instead of *drag_ref* and is divided by 4 to get the load per wheel (remember that the downforce is also equally distributed in all the wheels).

Straight Line:
Acceleration:

For simplicity assume that the acceleration is only limited by engine power and not by traction.

Longitudinal acceleration, a_x is:

$$a_x = \frac{\dfrac{P}{V} - Drag}{m} \qquad [m/s^2]$$

.... (14)

Where *Drag* is the vehicle drag in straight line.
Note: remember that engine power must be in watts.

For simplicity, assume that there is no drag from tire slip. Thus the drag in a straight line is the same as D_{total} in equation (2) without the term with slip angle, (equation (4)).

$$Drag = \frac{V^2}{V\_ref^2} \times drag\_ref \times 1000 + C_r \times m \times g$$

[N] .... (15)

As the rate of acceleration changes continuously, we can approximate the result by dividing the velocity in small increments of displacement at which the acceleration (a_x) is assumed constant. See the graph below:

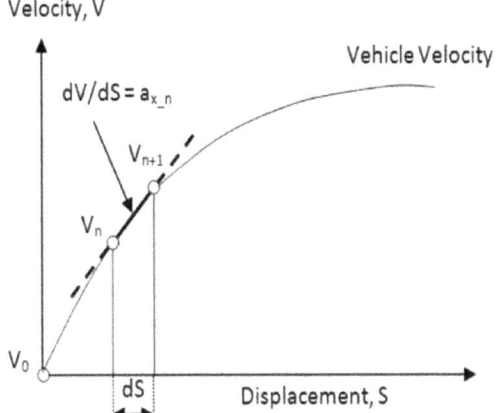

Velocity, V

Vehicle Velocity

$dV/dS = a_{x\_n}$

V_{n+1}

V_n

V_0

dS

Displacement, S

V_0 is the cornering speed.
The velocity is calculated starting from cornering speed where:

$$V_{n+1} = V_n + a_{x\_n} \times dt$$

[m/s] .... (16)

Where dt is the time split and can be calculated as follows:

$$dt = dS / Vn$$

This assumes that for a given time interval, the vehicle is travelling that distance at constant speed which is not true but if we divide the displacement in small increments (e.g. 5m for straights that are 500m long) we get only a small error associated with it and for this exercise this approximation is good enough.

> Note: The important part is that we are aware of all the assumptions we make along the way and understand the risks associated with it.

Braking:
In this case we assume that braking is predominantly determined by drag and tire grip and that longitudinal tire grip is proportional to vertical load. Also static loads and load transfer effects on tire vertical load were ignored for simplicity.

$$braking = \frac{\left[\left(\frac{V}{V\_ref}\right)^2 \times downforce \times 1000 + m.g\right] \times \mu_{tyre} + Drag}{m}$$

$[m/s^2]$.... (17)

μ_{tyre} is the friction coefficient for the tire when braking (use $\mu_{tyre}= 1.4$).
Drag from equation (15).

The braking velocity can be calculated in a similar way as for acceleration. In Excel a method to determine the braking point is to start from the following corner velocity and work backwards.

The displacement point where both acceleration and braking velocities meet is the braking point:

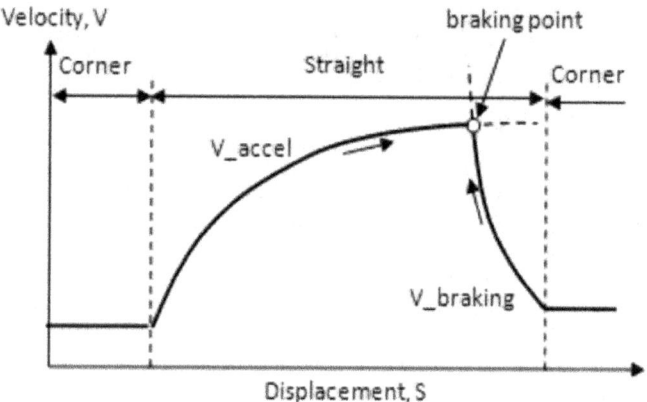

Programming this Lap Time in Excel, we get a circuit composed of 3 straights and 3 corners, we can change at will; running it, we get the total time and maximum speeds at every corner; times and speeds.

Vehicle mass, m	650	Kg
Engine power, P	600	Kw
CG Height, h	0,27	m
Track Width, track	1,5	m
DownForce, downforce	8,5	KN
Aero Eff., aero_eff	3,5	:1
Drag Reference, drag_ref	2,4	KN
g	9,8	m/s^2
Rolling res., cr	0,02	-
Slip angle, slip_a	6,0	º
Reference Velocity, v_ref	70	m/s

CORNER RADIUS	Corner 1, _r1	130		m
	Corner 2, _r2	190		m
	Corner 3, _r3	250		m
DISPL. IN CORNERS	Displ. 1, d_1	272		m
	Displ. 2, d_2	398		m
	Displ. 3, d_3	524		m
STRAIGHT LINE	Straight 1, straight_1	500		m
	Straight 2, straight_2	500		m
	Straight 3, straight_3	500		m
Corner 1	59,9	m/s		
Corner 2	76,9	m/s		
Corner 3	89,1	m/s		
Lap Time		32,68	seg	

We can extend the theory of a Lap Time simulator and take into account other factors making it much more complete and accurate:

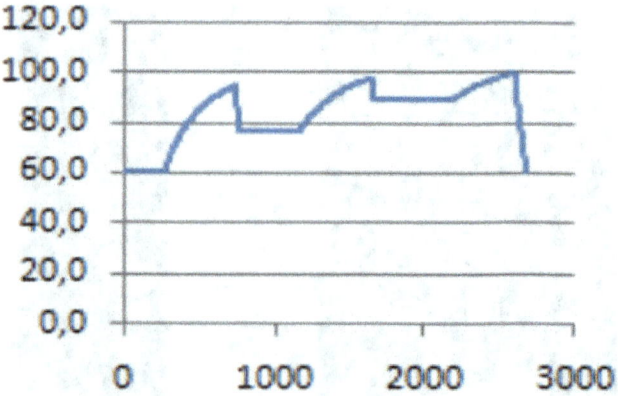

- Ratios.
- Engine curves.
- CG.
- CDP.
- Weights and Measures of the car. Suspended and non-suspended masses.
- Parameters or model tire; variation of pressure, temperature and geometry.
- Differential.
- Power loss in transmission.
- Percentage of braking.
- Front, rear or all-wheel drive.
- Ambient temperature and density.
- Diameter and general measures of the wheels.
- Lift and Drag.
- Fuel consumption.
- Etc.

One of the important factors to enter, is the modeling of tires; then transcribed in Matlab software to calculate based on the parameters or values Pacejka curves of lateral grip:

+ Name: PacejkaFy:
function [fy0, ay, K] = PacejkaFy (Fz, gy)
% Lateral force Fy calculated using the model Pacejka
%
% Nominal vertical force
Fnom = 2378.134018;
%

```
% coefficients
pdy1=-1.1368650;
pdy2=0.4689010;
pdy3=-5.2737330;
pcy1=0.9539170;
pey1=-0.2999010;
pey2=0.8782960;
pey3=-0.2283970;
pey4=-7.3951333;
phy1=0.0045202;
phy2=-0.0007918;
phy3=0.0438000;
pky1=-11.8621944;
pky2=1.2972245;
pky3=0.1856351;
pvy1=0.1338810;
pvy2=-0.1655041;
pvy3=-0.0526262;
pvy4=-0.8147694;
%

% Scale Factors
lFz0=1;
lCy=1;
lHy=1;
lEy=1;
lKy=1;
lVy=1;
lmy=1;
%
Fz0=FNOM*lFz0;
dfz=(Fz-Fz0)/Fz0;
my=(pdy1+pdy2*dfz)*(1-pdy3*gy^2)*lmy;
D=my*Fz;
C=pcy1*lCy;
SH=(phy1+(phy2*dfz)+(phy3*gy))*lHy;
SH0=(phy1+(phy2*dfz))*lHy;
%SH0=(phy1+(phy2*dfz))*lHy;
c=2*atan(Fz/(pky2*Fz0));
K=pky1*Fz0*lKy*(1-pky3*abs(gy))*sin(c);
B=K/(C*D);
SVy=Fz*(pvy1+(pvy2*dfz)+(pvy3+(pvy4*dfz)*gy))*lVy
*lmy;
SVy0=Fz*(pvy1+(pvy2*dfz)+pvy3)*lVy*lmy;
%
% Range of slip angles
```

```
ext=(9*pi)/(180);
extinf=-ext;
extsup=ext;
incr=(0.5*pi)/(180);
    cont=0;
for b=extinf:incr:extsup
    i=cont+1;
    a(i)=b;
    ay(i)=(a(i)+SH-SH0);
    ay';
    E=(pey1+(pey2*dfz))*(1-
(pey3+(pey4*gy))*sin(ay))*IEy;
    %Calculo de la fuerza lateral
    Fy0(i)=D*sin(C*atan(B*ay(i)-E(i)*(B*ay(i)-
atan(B*a(i)))))+SVy-SVy0;
    cont=cont+1;
end
end
□ Name: Pacejka
clear
% Pacejka model for the lateral force
% Menu to select the camber to represent graphically
mypick = menu ('Choose camber angle for Fy vs. slip
curve represent', '0', '0.5' '1' '1.5', 'two', '2.5', '3', '3.5' ,
'All')
switch mypick;
    case 1
        opt1=0;
        opt2=opt1;
    case 2
        opt1=0.5;
        opt2=opt1;
    case 3
```

```matlab
            opt1=1;
            opt2=opt1;
        case 4
            opt1=1.5;
            opt2=opt1;
        case 5
            opt1=2;
            opt2=opt1;
        case 6


            opt1=2.5;
            opt2=opt1;
        case 7
            opt1=3;
            opt2=opt1;
        case 8
            opt1=3.5;
            opt2=opt1;
        case 9
            opt1=0;
            opt2=3.5;
end
%
for Fzkg=75:75:150
    Fz=9.81*Fzkg;
for gydeg=opt1:0.5:opt2
    gy=(gydeg*pi)/180;

% Function call
[Fy0, ay, K] = PacejkaFy (Fz, g);
% Calculation of the area under the curve (information)
area = trapz (fy0, ay);
% Cderiva = K / Fz;
% By screen printing
fprintf ('For a vertical load% d \ n', Fzkg)
fprintf ('Angle drop% d \ n', gydeg)
fprintf ('stiffness coefficient derived% d \ n', K);
fprintf ('Area under the curve% d \ n ", area);
fprintf ('\ n');
% Representation of the curves plotted
% Fz '
    %Fy0'
    %(Fy0/Fz)'
    plot(ay,Fy0);
```

```
    grid on;
    hold on;
  end
end
%
```

The program asks you a series of values and represents the desired graph.

For a vertical load 75
Angle 1
Stiffness coefficient −1.269040e+05
Area below curve 3.304372E-01
 For a vertical load 150
Angle 1
Stiffness coefficient −2.185257e+04

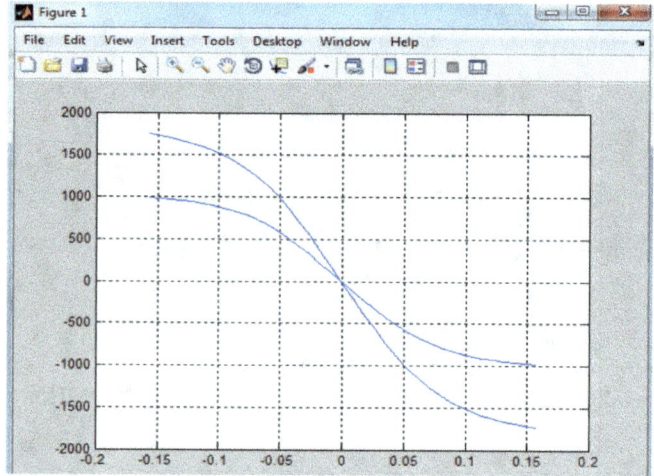

A Lap Time simulator is not used to quantify times, speeds or consumption (although very accurately quantified until the extent that you can plan race strategies), but basically it is used to compare different setup options. In the previous "expanded" lap time, we can generate several macros for the "optimal" values that make the minimum lap time; we can vary the values from a number to another number, a step (interval) given and know the optimal value within that range:

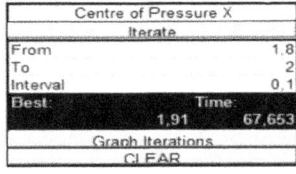

Centre of Pressure X		
Iterate		
From		1,8
To		2
Interval		0,1
Best	Time:	
	1,91	67,653
Graph Iterations		
CLEAR		

Wheelbase		
Iterate		
From		2,7
To		3,3
Interval		0,1
Best	Time:	
	3,1	67,100
Graph Iterations		
CLEAR		

Thus we know the optimal values for each variable. The problem we face is to establish a protocol or study strategy to analyze optimal values of sets of variables; it is a fascinating subject of mathematical study.

With this type of calculation tools, it is possible, from a strictly aerodynamics point of view to:

- Know if it's important or not and to what extent, the downforce required for a given circuit. There will be circuits, e.g. Monaco, where downforce is very important and others, in which it will hardly be.
- Understand the relationship between downforce and drag, also called aero efficiency for optimum efficiency is also important: not because we generate more downforce that the total time will be less:

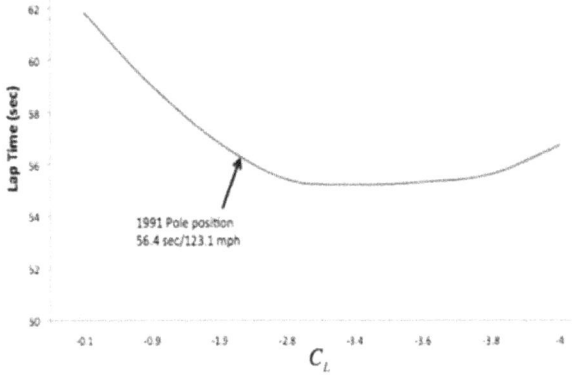

$$C_D = 0.04(C_L - 0.2)^2 + 0.3$$

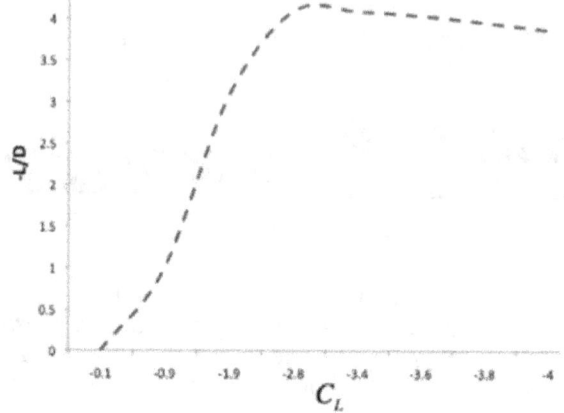

There are other factors that determine the efficiency of the car in relation to lap time:

1) It's called Lap-Time coefficient:

$$\Delta Lt = \frac{\delta Lt}{\delta Cz}\Delta Cz + \frac{\delta Lt}{\delta Cx}\Delta Cx$$

Looking at each of the coefficients, we realize what this overall coefficient of lap-Time is:

It's the function of downforce or load variation with respect to the variation of load factor and the drag coefficient.

$$\frac{\delta Lt}{\delta Cz}/\frac{\delta Lt}{\delta Cx}$$

This value is close to 1 for low speed circuits and high loads.
It's close to 2 for medium load circuits. It is close to 3 for low load circuits

2) There is another factor called "Lap Time Index". It indicates that it is not enough to produce more downforce or enlarge the aerodynamic efficiency to get less time per lap; you need to improve these parameters in a smart way.

$$LM_{INDEX} =- 100 \cdot \left[\frac{\delta T\%}{\delta Cz} \cdot (Czt_{REB} - 2) + \frac{\delta T\%}{\delta Cx} \cdot (Cxt_{REB} - 0.5) \right]$$

Where:

$Czt = 2$
$Cx = 0.5$
$\bar{B}al\% = 45\%$

Sensitivities or partial derivatives are calculated either by track data or from simulation software.
Other important values that define the dynamics of the car are:
3) Pitch sensitivity:

$$\partial Cz_F / \partial H_R$$

4) Frontal area sensitivity:
$$\partial Cz_F / \partial H_F$$

These values define the sensitivity of the total downforce, depending on the variations of the front wing and the front height: everything that happens at the front strongly influences the rear part.

Finally, a tire test would be appropriate for lap times in terms of tire degradation and more: knowing the time of the pit stop, knowing full-time races:

We know that the lap times of each car changes according to the percentage of race completed; Cars often reduce lap times because they are losing mass due to fuel consumption. But there is a time when the tire loses are a fundamental characteristics and time increases; it is just the moment to change tires.

We have seen before how to make a lap time in a simple way, simulating the car fully. But to finish the track problem completely, it is necessary to introduce two improvements to simulate races and strategies:

• The mass change every lap, subtracting

the mass of fuel consumed in the previous round.

- The tire degradation.

These two additions are essential and very important.

We know that modeling tires is a function of temperature, pressure and several other condition, but in short, combining all these factors we arrive at friction graphs depending on the number of laps completed; we can have tires that last many turns and others, that last few laps; these graphs are generated for each race and are susceptible to many variations depending on temperature and other variables; but still make a big "help":

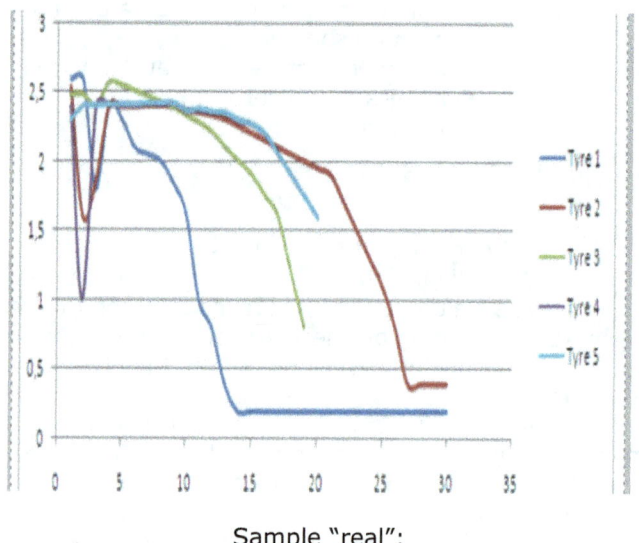

Sample "real":

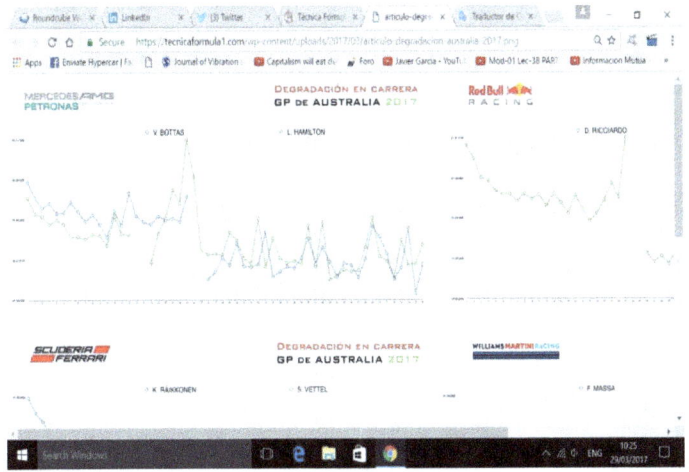

We can choose a tire or another, depending on many variables.

When should we change a tire for another? When lap times aren't reduced.

As if that was not enough, we can set a pit stop time for tire change, so we have a full race simulation in the program automatically changing each tire in the order you want:

Track	Top Gear test track		Select
Laps	7	Distance	20.2321 Km
Fuel Load	35	Kg	
Box Time	22	secs	
Tyre Change Criteria		104 %	
Run			

A simple example is explained below applying everything we have learn up to now:

⇒ Before a rally race, like the one in Arrate in the Basque Country (Spain), we need to know whether the aero is important or not; i.e.: if we put downforce load, how much downforce is generated and what is the time gain by placing wings? They are very basic but extremely important questions.

We apply our lap time in Excel.

Through google earth, we can know the path of the race:

		Altitude:
Start GI-3950	km.: 0.730	195m
		Altitude:
Finish GI-3301	km.: 1.750	570m

Longitude: 4990m	**Average Slope:** 7.51%
Height Difference: 375m	**Maximum Slope:** 11.4%

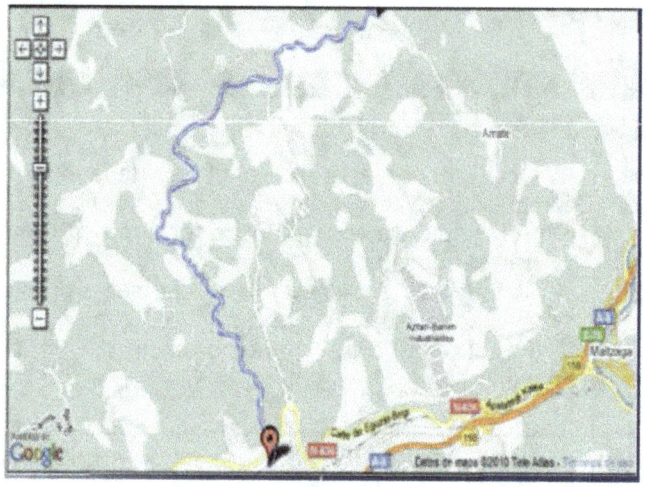

We analyze 3 different setups.

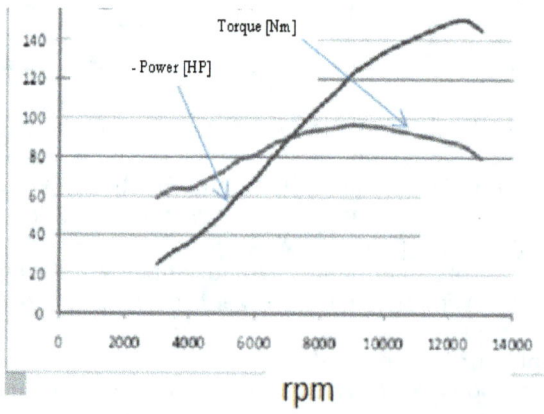

Chassis			Variable
mass	550	[kg]	m
Weight split	40	[%]	Wf
CG height	0.35	[m]	h
Wheelbase	2.39	[m]	L
Track	1.55	[m]	T
Susp_stiff. Split	40	[%]	K_f
Brake split	68	[%]	B_f
Max brake capacity	4	[g]	B_max
drag coeff.	0.4	-	Cd
lift coeff.	0	-	Cl
Area	1.5	[m^2]	A_
Centre of pressure X	1.434	[m]	CPx
Centre of pressure Z		[m]	CPz

Tyres

coefficients:	Front	Rear	
Lat a1_	-52	-48	a1_f, a1_r
Lat a2_	1058	1141	a2_f, a2_r
Long b1_	-21	-19	b1_f, b1_r
Long b2_	1164	1255	b2_f, b2_r

Transmission

Primary drive	1.717	:1	PR
1st	2.286	:1	G1_
2nd	1.778	:1	G2_
3rd	1.500	:1	G3_
4th	1.333	:1	G4_
5th	1.214	:1	G5_
6th	1.138	:1	G6_
Final drive	2.975	:1	AR
Diff TBR	2.5	:1	TBR
Wheel radius	0.285	m	r
Driveline loss	18	%	drive_loss

Engine Speed [RPM]	Torque [Nm]	Power [BH
3000	59	25
3500	64	31
4000	64	36
4500	68	43
5000	72	51
5500	79	61
6000	81	68
6500	86	79
7000	90	89
7500	93	98
8000	94	106
8500	95	114
9000	97	123
9500	96	128
10000	95	134
10500	94	138
11000	92	142
11500	90	145
12000	88	149
12500	86	150
13000	79	145

	SETUP 1	SETUP 2	SETUP 3
TIME	02:49,1	02:46,0	02:46,4
Cd	0,4	0,6	1,2
Cl	0	-0,3	-0,6

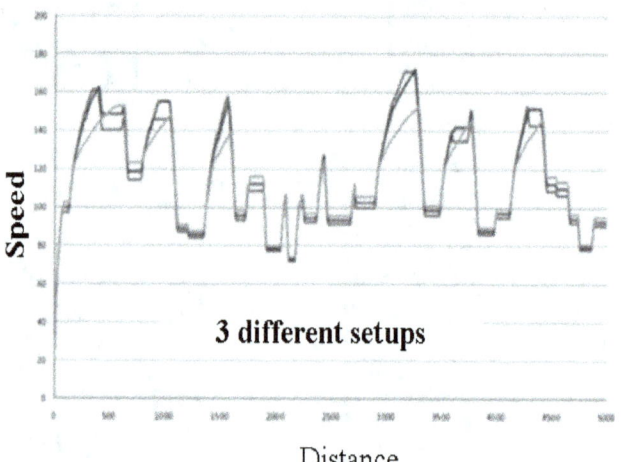

3 different setups

Distance

Result: having more downforce means a reduction in lap time. It is a nice, easy and simple example of a direct application to a normal and everyday questions; using and generating lap times, we can determine what might be the ideal load for a given setup.

Article Willem Toet about that:
Aerodynamic forces are the number 1 performance-determining and differentiating factor in Formula 1 as well as in many branches of motorsport involving open-wheeled and sports cars. That's a pretty bold statement with the present state of powertrains in F1 but analysis says it holds true, despite big differences in powertrains. There is still more lap time difference from aero than from power. Red Bull have the problem that they are fighting other teams that also have great aerodynamics.
The plot shows speed (vertical scale) against distance (horizontal scale) for the 2012 F1 car I know well, along with several theoretical cars. The present generation of cars will have similar performance characteristics (but I am not permitted to show this data).

We all know that, speaking generally, the forces exerted by aerodynamics increase with the square of speed. Perhaps that is why I am often asked "From what speed does aerodynamics make a difference". I think from the speed of a tight hairpin bend – in other words from nearly nothing. You won't get a big percent speed improvement in slow corners, but you spend "ages" waiting to be able to accelerate so even a tiny improvement will make lap time. If you look at the plot showing speed against distance, perhaps it isn't so clear. So I've created a plot of the same data showing speed against time (two traces only). The problem with a time plot is that the corners move to different locations so it is harder to compare. However, if you focus on the area around 26 seconds where the "cars" are aligned in time, you can see how much more time it takes the car without downforce to cover this part of the lap. So the long duration of slow corners means that aero is important there too.

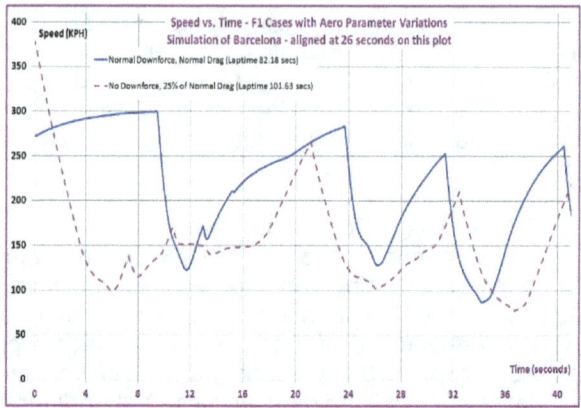

Speed against time.   Much harder to follow because the places on the track move about on the plot but it should be clear that the car without downforce (which also has very low drag) takes much more time to get around the corner centred around 26 seconds than the car with downforce.

The other corners would show the same trend if they were aligned.

Back to the first plot, which shows how downforce and drag can influence the limit speed of a car over part of a lap of the Barcelona track.   Zero downforce is certainly achievable but getting the drag down to 25% of today's real car values is probably not possible for a legal F1 car.   No problem - what we are trying to look at here is the limit of performance.   That's why I've chosen such an extreme drag reduction (to give drag the best chance I can to influence performance).

The curves are created using lap-time simulation software, which is regularly validated against real-car performance.   It is assumed that other parameters such as tyres are at the same performance level, which is not strictly realistic but that is not that important for the purposes of this discussion.   Removing downforce increases lap time by more than 20 seconds.

Reducing the drag of the car to just 25% of its real value, gives much less than 5 seconds of lap time gain, and just 2 seconds if you have no downforce.   Why does the gain vary due to downforce level?

If you are at the cornering limit because you have no downforce, a drag reduction is not going to help much until you get to a straight and only then once you have enough grip to use all that power. Alternatively, if you can fly around the corner because you have grip, your average speed is higher and you will have to 'fight' drag more, using power – so a drag reduction will help you more in those circumstances. In low-speed corners little power is needed to maintain speed, so reducing the drag has almost no effect.

In high-speed corners the influence of drag can become quite significant. However, it makes 10% of the difference you make with downforce and that is with lower drag than can actually be achieved.

	Corner type	Speed without downforce	Speed with normal downforce	Speed increase due to downforce	Speed increase possible in corners due to drag reduction to 25% of normal value
A	high-speed	139.5 kph	235.5 kph	69%	7.0%
B	medium-speed	101.4 kph	127.6 kph	25%	0.34%
C	low-speed	76.6 kph	86.2 kph	12%	0.2%

Table 1: Speed variables depending on corner speed

In extreme cases, e.g. when all cars are easy flat though a given (very high-speed corner), then the situation changes and, practically speaking, only drag will be important. Teams take this into consideration when selecting the downforce level for a given race track as the ideal compromise changes.

The designs of the race tracks in many ways drive the compromise teams use for selecting the drag level of the car used for their development work.

Sergio Perez in the 2012 Sauber Ferrari F1 car – I am not sure who to thank for the picture (Sauber Motorsport I believe).

The designs of the race tracks in many ways drive the compromise teams use for selecting the drag level of the car used for their development work. Modern tracks tend to be centred around a restricted spectator area. This means they have more corners and shorter straights. That drives designs in the more extreme direction of downforce (rather than worrying about drag).

Looking at today's Formula 1, it is clear that both powertrain and aerodynamics are critical to performance. For a privateer team, modifying the engine is not realistically an option. I say realistically also because the engine would need to be re-badged.

This is something that only Red Bull could afford to do. For a factory team, both clearly need to be at the highest level. However, recovering from a deficit in powertrain will cost more than recovering from an aerodynamic deficit.
P.S.

Here's an alternative way to see how time plots stretch low speed corners. The same car plotted from the same start and finish places on the track. Same data.

The two lines are speed vs. distance and speed vs. time. Time squeezes the high speed and stretches the low speed. Lap TIME is important....

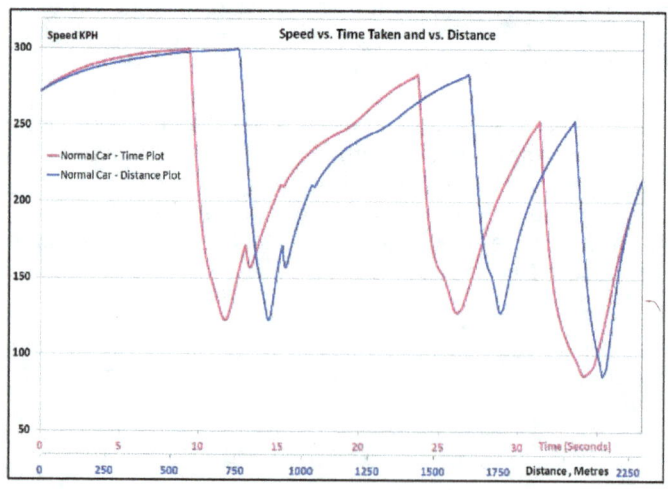

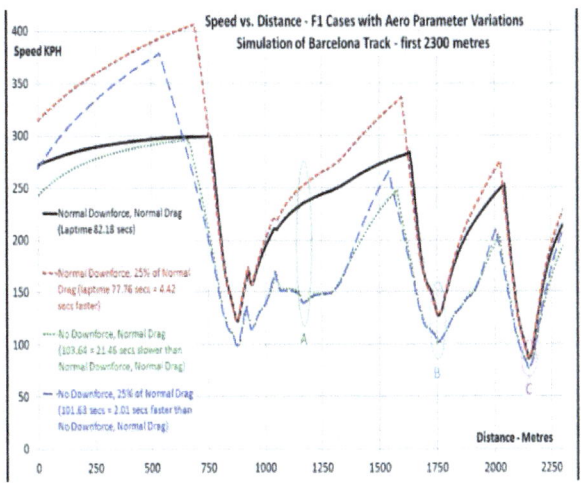

Speed vs. Distance - F1 Cases with Aero Parameter Variations
Simulation of Barcelona Track - first 2300 metres

Is possible use this lap time type, by with rain…
how ??

- Reducing the friction tire.
- Reducing engine power….

CORNER TOP SPEED

The maximum speed through a corner is the maximum speed without losing traction on any wheel. This speed depends on many factors, and we will show an aero balance effects study.

To study how the aerodynamic balance affects cornering, we conducted a quasi-static simulation in which the car is traveling around a curve of constant radius at different speeds.

For each speed we calculate a lateral acceleration and an aerodynamic force. From these and model of vehicle roll is determined how much mass is transferred on each shaft as a function of the ratio of roll stiffness and roll shaft position. Once known how much each wheel is loaded, using Pacejka model, the characteristic curve of each tire is calculated. Finally, through the requirement of lateral force on each axis, drift angles for each wheel and a budget of understeer is determined. This budget of understeer characterizes the behavior of the vehicle.

Is necessary to know if the car have understeer or oversteer; that in function of aerodynamic balance or center pressure position; for that, is possible to study it; we will see this effect in another chapter.

AERODYNAMIC DRAG

The other fundamental force in competition is aerodynamic drag; while downforce served to stick the car to the road, lower drag means:

- Increased acceleration.
- A higher top speed.
- Reduction consume fuel. About that:

The energy content of gasoline is about 32 x 106 J / litre, but because of the engine efficiency only 25% of this chemical energy gets converted to mechanical energy.

So what about Air Resistance?

When we drive a car we leave behind us a big tube of air that is swirling around (See Figure 1). The passage of the car is what makes the air swirl around, so our car engine needs to provide all the energy for all of that swirling. Figuring out all the details of exactly which air is swirling where is not important; we just want to make a reasonably accurate estimate of how much energy this will cost us, so we'll develop the following model.

The swirling air is confined to some region near the path of the car. Let's imagine this region is a long tube, with a cross sectional area Atube, and that the passage of the car makes it swirl with velocity v, which is the same velocity as the car. The area Atube is similar to the frontal area of the car, but not exactly the same. A more streamlined car will have Atube slightly smaller than the frontal area of the car. The ratio of Atube / Acar is called the Drag Coefficient (CD). For a typical family sedan, CD= 0.33 and for a cyclist, CD= 0.93

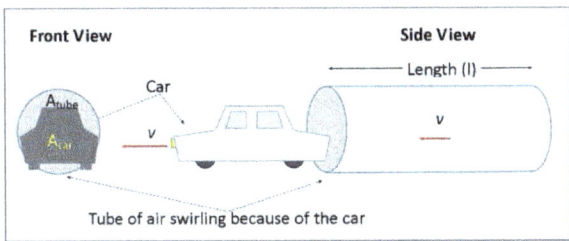

Wikipedia.

Tube of air swirling around a moving car.

We want use this idea to figure out how much energy it costs the car per kilometre travelled. We can figure out how much energy the car loses to the air by figuring out the kinetic energy of this tube of moving air. To figure out kinetic energy we just need the mass and the volume of the tube of air. A car travelling at speed v will also make the air travel at speed v, so all we need to do is get the mass. Say the car travels for some distance d. The length of the tube of air that the car encounters in that distance will be the same d:

$$\bullet \; \text{Length} = d$$

So the total volume of this tube will be:

$$\bullet \; \text{Volume} = (\text{Area})(\text{Length}) = A_{tube}d$$

And the mass of the tube will be:

$$\bullet \; \text{Mass} = (\text{Density})(\text{Volume}) = \rho A_{tube}d$$

So now the kinetic energy of the tube will be:

$$KE \; = \; \frac{1}{2}mv^2$$

$$= \; \frac{1}{2}A_{tube}dv^2$$

$$= \; \frac{1}{2}\rho A_{car}C_D dv^2$$

Given that the area of a typical family sedan is

$$A = (2 \text{ m})(1.5 \text{ m}) = 3 \text{ m}^2$$

let's see how much work is done against air resistance for each kilometre a typical car driving at 50 km/h (14 m/s) travels.

$$\text{Work done against air resistance} \; = \; \frac{1}{2}\rho A_{car}C_D dv^2$$

$$= \; \frac{1}{2}(1.3 \text{ kg/m}^3)(3 \text{ m}^2)(0.33)(1000 \text{ m})(14\text{m/s})^2$$

$$= \; 126,126\text{kg} \bullet \text{m}^2/\text{s}^2$$

$$= \; 126 \text{ kJ}$$

So, for each kilometre travelled, 126 kJ of work is done against air resistance.

We can figure out how much fuel is required for each kilometre travelled using the efficiency formula:

$$\text{Efficiency} \; = \; \frac{\text{Work Output}}{\text{Work Input}}$$

$$= \; \frac{\text{Work Output}}{\text{Fuel Energy Input}}$$

$$\text{Fuel Energy Input} \; = \; \frac{\text{Work Output}}{\text{Efficiency}}$$

$$= \; \frac{126 \text{ kJ}}{25\%}$$

$$= \; 505 \text{ kJ}$$

And to provide this amount of energy we need to use

$$\text{Efficiency} \quad = \quad \frac{\text{Work Output}}{\text{Work Input}}$$

$$= \quad \frac{\text{Work Output}}{\text{Fuel Energy Input}}$$

$$\text{Fuel Energy Input} \quad = \quad \frac{\text{Work Output}}{\text{Efficiency}}$$

$$= \quad \frac{126 \text{ kJ}}{25\%}$$

$$= \quad 505 \text{ kJ}$$

So, 0.016 L of fuel is required to drive 1 km.

If we compare this with our earlier rule of thumb that the typical fuel consumption of a car is 0.076 L/km4. We see that air resistance is only accounting for 21% of the energy cost. This is because we did the calculation at 50 km/h. At this speed, air friction is really a very small part of the fuel requirements of a car, which is why sometimes we choose to neglect it in our calculations. However, because the fuel consumption depends on the velocity squared, air resistance becomes much more important at higher speeds.

At 100 km/h, the fuel consumption will be FOUR times higher, or 0.064 L/km. This is much closer to 0.076 L/km. To get an even better understanding of energy consumption in cars, we can also take into account the rolling resistance of the car.

Remember that we are considering only the energy needed to keep the vehicle moving at a constant speed. Most automobiles have many other systems that consume fuel as well (e.g. drive-train losses, standby, accessories such as air conditioning). Further, engine efficiency will not be a constant in reality, but rather optimized for certain speeds. Nevertheless, our calculations provide a very useful lower bound for fuel economy.

Summary:

Can we use these ideas to figure out how to get more mileage out of our cars? Well it looks like when you're travelling at high speed MOST of the gasoline goes into making the air swirl around.

There are several types of drag forces:

- Parasitic drag consisting of:
  - Form drag
  - Skin friction named also viscous
- Molecules impact (named also pressure force).
- Separation boundary layer or flow (turbulent flow)
- Interference drag
- Lift-induced drag
- Wave drag (hydrodynamics)

About the pressure and viscous drag (Wikipedia images):

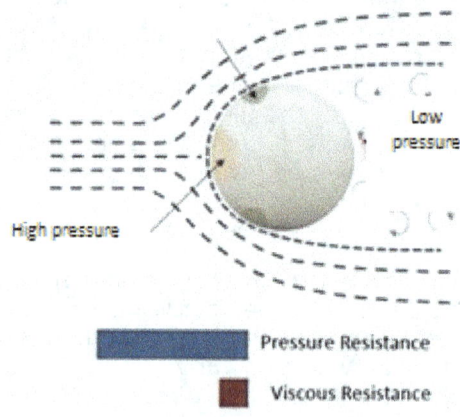

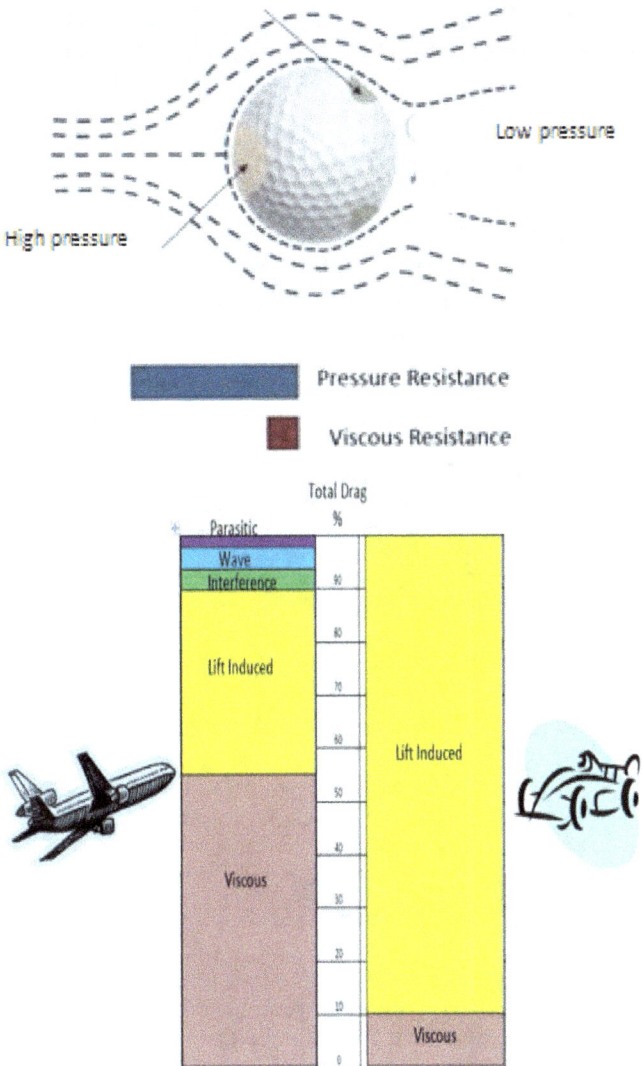

High pressure

Low pressure

Pressure Resistance

Viscous Resistance

Total Drag %

Parasitic
Wave
Interference
Lift Induced
Viscous

90
80
70
60
50
40
30
20
10
0

Lift Induced
Viscous

The turbulence, is the "flow or condition", that produces almost all the drag; is possible reduce the turbulence? Yes:

A good representation artistic of drag:

⇒ The induced drag as the name suggests is induced by another force, such as downforce (L). We can easily calculate this drag with the following equation:

$$Di = \frac{2 \cdot L^2}{\rho \pi b^2 V^2 e}$$

Where "b" is the wingspan, "V" is the wing speed and "e" is a parameter that depends on the shape of the wing (elliptical, rectangular, plant, etc...). Do not forget that these values are vector. Horizontal drag does not opposite to lift vertical, however itself opposes thrust. By this I mean that being in orthogonal planes, cannot be operated as absolute values:

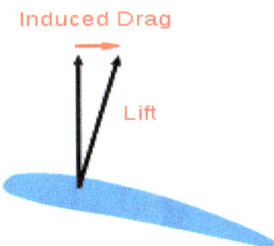

The induced drag increases with the square of downforce, but decreases with the square of the wingspan; hence, sailplanes, have long narrow wings (high aspect ratio) in order to mitigate or reduce its lift-induced resistance; note: the speed exist in "L"....

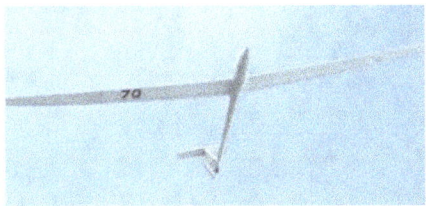

Exist another expression for calculating this drag:

$$C_D = C_{Do} + \frac{k\,C_L^{2}}{\pi A}$$

D_{Do} is drag for lift zero; "A" is wing aspect ratio (b/c); k=e; c=chord=area/span

Induced Drag Coefficient

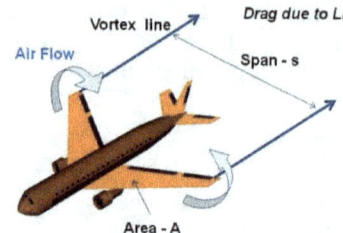

Drag due to Lift

Air Flow

Vortex line

Span - s

Area - A

Aspect Ratio = AR

$$AR = \frac{s^2}{A}$$

$$Cd_i = \frac{Cl^2}{\pi \, AR \, e}$$

Efficiency factor = e
For an ellipse, e = 1
In general e < 1

Pressure difference from top to bottom of the wing causes spillage around the wing tips.

Downwash from the tips induces local angle of attack with additional drag component on a finite wing.

Taper ratio = chord (end) / chord (start wing in body)

$$D_i = \frac{1}{2}\rho V^2 S C_{Di} = \frac{1}{2}\rho_0 V_e^2 S C_{Di}$$

where

$$C_{Di} = \frac{C_L^2}{\pi e AR} \quad \text{and}$$

$$C_L = \frac{L}{\frac{1}{2}\rho_0 V_e^2 S}$$

Thus

$$C_{Di} = \frac{L^2}{\frac{1}{4}\rho_0^2 V_e^4 S^2 \pi e AR}$$

Hence

$$D_i = \frac{L^2}{\frac{1}{2}\rho_0 V_e^2 S \pi e AR}$$

where:

AR is the aspect ratio,

C_{Di} is the induced drag coefficient (see Lifting-line theory),

C_L is the lift coefficient,

D_i is the induced drag,

e is the wing span efficiency value by which the induced drag exceeds that of an elliptical lift distribution, typically 0.85 to 0.95,

L is the lift,

S is the gross wing area: the product of the wing span and the Mean Aerodynamic Chord.[1][P

V is the true airspeed,

V_e is the equivalent airspeed,

ρ is the air density and

ρ_0 is 1.225 kg/m², the air density at sea level, ISA conditions.

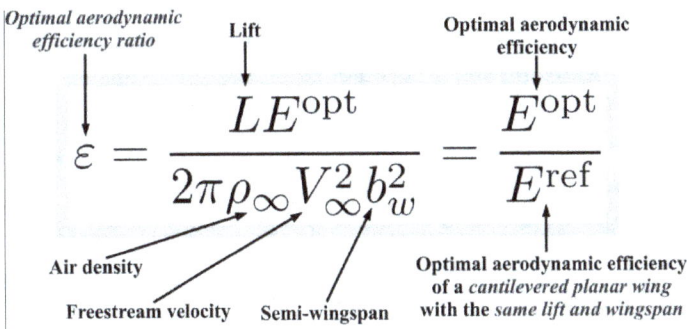

Optimal aerodynamic efficiency ratio

Lift

Optimal aerodynamic efficiency

$$\varepsilon = \frac{LE^{\text{opt}}}{2\pi\rho_\infty V_\infty^2 b_w^2} = \frac{E^{\text{opt}}}{E^{\text{ref}}}$$

Air density

Freestream velocity   Semi-wingspan

Optimal aerodynamic efficiency of a *cantilevered planar wing* with the *same lift and wingspan*

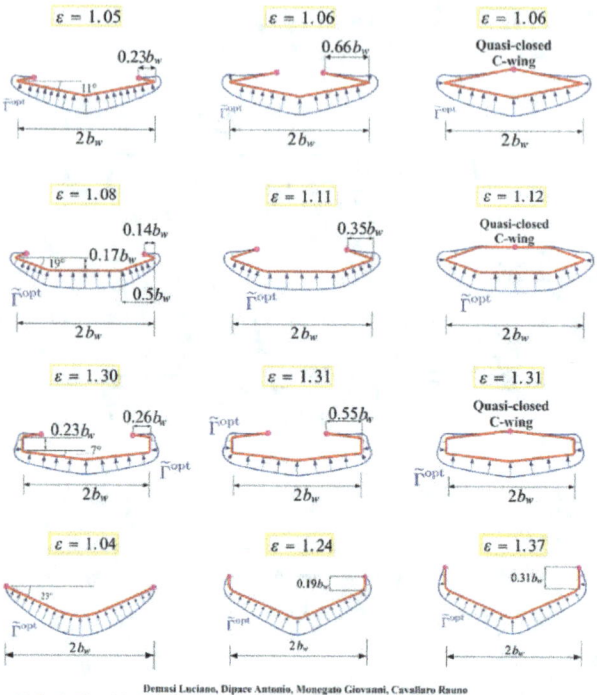

Demasi Luciano, Dipace Antonio, Monegato Giovanni, Cavallaro Rauno
"An Invariant Formulation for the Minimum Induced Drag Conditions of Non-planar Wing Systems" AIAA Journal, in press (2014)

Basically, the induced drag is created by the vortex extremals and turbulence
iiii

There is another type of induced drag which is called induced turbulence drag; This turbulence may cause vibrations and in fact it is generated in most cases.

These vibrations can cause structural damage (resonance); such was the case with the famous Tahoma Bridge:

On a smaller scale vibrations caused by turbulence may appear in circular steel chimneys. To prevent the formation of turbulence we can include some additional elements to the chimney:

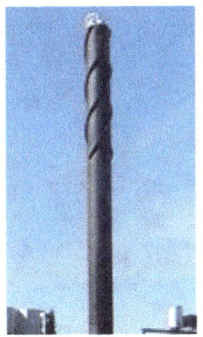

We can see these devices in the picture above.

Another think very important to designing one piece, is know his resonant frequencies:

Every parts in human body, have his self-frequency resonant; this frequency is very dangerous so is possible, for example, loss of sight if the frequency is around 20-80 Hz; that problem was occurred in Bilbao street circuit some years ago:

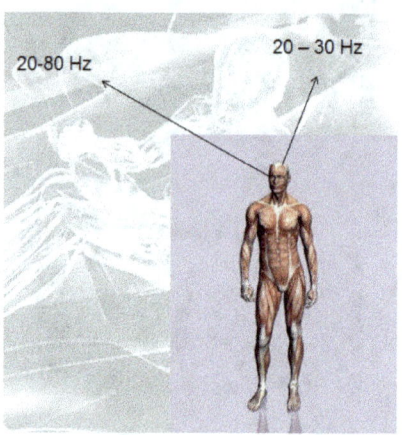

If the frequency is 1.2 Hz, the person gets dizzy (ships for example):

To calculate the parasitic + induced drag, we use the same expression we used to calculate downforce, but substituting the downforce coefficient "Cl" for the drag coefficient "Cd":

$$D = \frac{1}{2}\rho A V^2 Cd$$

We can classify vehicle's drag based on the part of the car where it is generated:

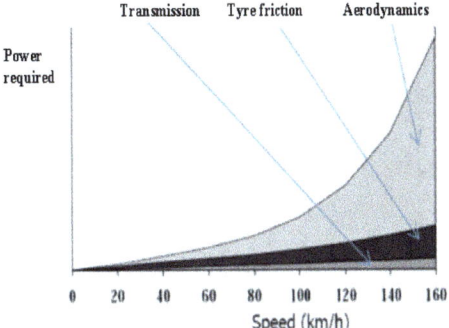

The most important one especially at "high" speed is aerodynamics.

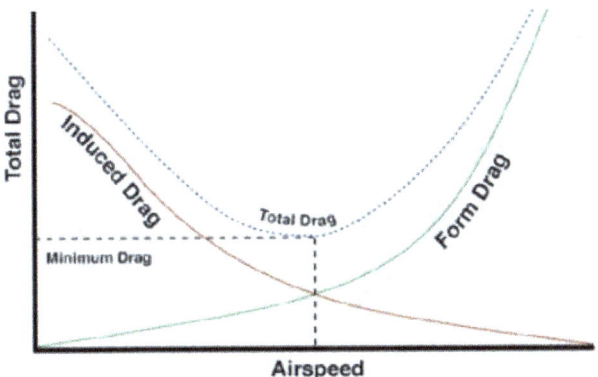

Also, the drag depend of Reynolds number:

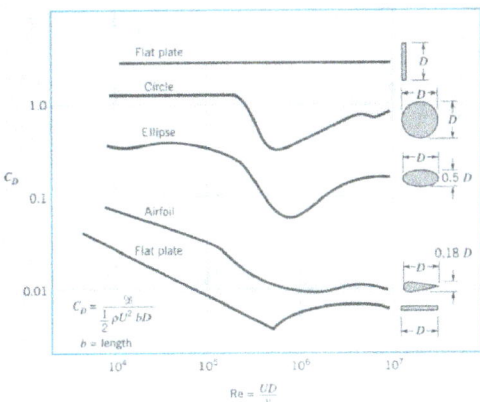

About the relation between drag and geometry, see the next images; first image: the geometry has more drag because the sumatory red vectors (drag pressure) are higher; the drag vectors have direction:

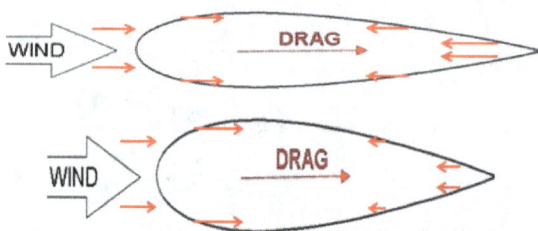

Furthermore, despite what many people believe, a racing car or high performance sports car, generates much more drag than a street car; This is due to the lift-induced drag: (very important: "Cx = Cd * A):

$Cx=0.31$

$Cx=0.31$

It is possible to have 2 cars with the same Cx (same as Cd); but the frontal areas can be different....

This is a very important fact to consider when comparing aerodynamic behavior of two different cars.

Here we can take a look at a small tables of drag values for different car types and shapes:

Mercury Topaz		0.36
Toyota Celica		0.34
Chevrolet Corvette		0.34
Dodge Daytona Turbo		0.34
Citröen		0.31
1932 Fiat Balilllo		0.60
Volkswagen "Bug"		0.46
Volkswagen Van		0.42
Volkswagen Scirocco		0.39

The shape of an object has a very great effect on the amount of drag.

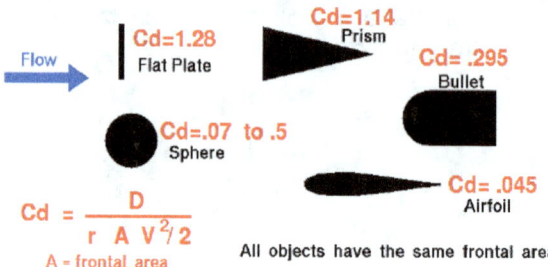

Flow

Cd=1.28
Flat Plate

Cd=1.14
Prism

Cd= .295
Bullet

Cd=.07 to .5
Sphere

Cd= .045
Airfoil

$$Cd = \frac{D}{r\ A\ V^2/2}$$

A = frontal area

All objects have the same frontal area.

AUDI	α	VW	α
A2	0.25-0.28	Lupo	0.29-0.33
A3	0.31	Polo	0.31-0.33
A4	0.29-0.34	Golf	0.32-0.35
A6	0.32-0.35	Bora	0.31-0.33
A8	0.29-0.30	Passat	0.27-0.32
TT	0.32-0.35	Phaeton	0.31
		New Beetle	0.37-0.38

AUDI	α	VW	α
A2	0.25-0.28	Lupo	0.29-0.33
A3	0.31	Polo	0.31-0.33
A4	0.29-0.34	Golf	0.32-0.35
A6	0.32-0.35	Bora	0.31-0.33
A8	0.29-0.30	Passat	0.27-0.32
TT	0.32-0.35	Phaeton	0.31
		New Beetle	0.37-0.38

ALFA ROMEO	α		BMW	α
147 Twin Spark	0.34		316	0.32
156	0.29-0.34		318	0.29
166	0.30		320	0.29-0.32
GTV	0.35		328	0.30-0.31
Spider	0.38		330	0.30-0.31
			325	0.35
			528	0.30-0.31
			745	0.29
			Z3	0.41-0.43
			M5	0.31
			Z8	0.41

SEAT	α		SKODA	α
Arosa	0.32-0.33		Fabia	0.29-0.34
Ibiza	0.30-0.31		Octavia	0.29-0.32
Cordoba	0.30-0.31		Superb	0.30-0.32
León	0.32-0.34			
Toledo	0.31			

CITRÖEN	cx		FIAT	cx
Saxo	0.35		Panda	0.42
C3	0.34		Seicento	0.36
Xantia	0.33-0.34		Punto	0.35-0.42
XSara	0.32-0.34		Stilo	0.32-0.34
C5	0.32-0.33		Marea	0.33-0.35
			Brava	0.34
			Palio	0.36
			Multipla	0.32
			Coupe	0.36
			Barchetta	0.44

FORD	cx		NISSAN	cx		OPEL	cx		PEUGEOT	cx
Fiesta	0.34		Micra	0.39		Agila	0.37		106	0.36-0.38
KA	0.35		Primera	0.36-0.37		Corsa	0.33-0.40		206	0.34-0.37
Focus	0.31-0.33		Almera	0.35		Astra	0.30-0.34		306	0.37
Mondeo	0.31-0.33					Vectra	0.30-0.33		307	0.32
Puma	0.35					Omega	0.32		406	0.31-0.33
Cougar	0.32					Zafira	0.32-0.33		607	0.29
						Tigra	0.34			
						Speddster	0.38			

RENAULT	cx		SUZUKI	cx		TOYOTA	cx		VOLVO	cx
Twingo	0.37		Baleno	0.40		Yaris	0.33-0.34		V40	0.34
Clio	0.36-0.37		Alto	0.40		Corolla	0.36-0.38		C70	0.32
Megane	0.34-0.40		Swift	0.39		Celica	0.37		S60	0.30
Avantime	0.36					Lexus	0.29-0.33		S70	0.34
Laguna	0.34								V70	0.33
Vel Satis	0.34								S80	0.31

MINI	cx		MERCEDES	cx		FERRARI	cx		PORSCHE	cx
One	0.37		A	0.31-0.33		355 F1	0.31		911 Carrera	0.31

Flag fluttering:

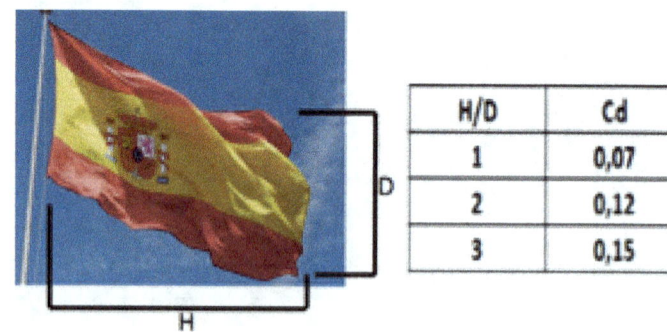

H/D	Cd
1	0,07
2	0,12
3	0,15

Different shapes:

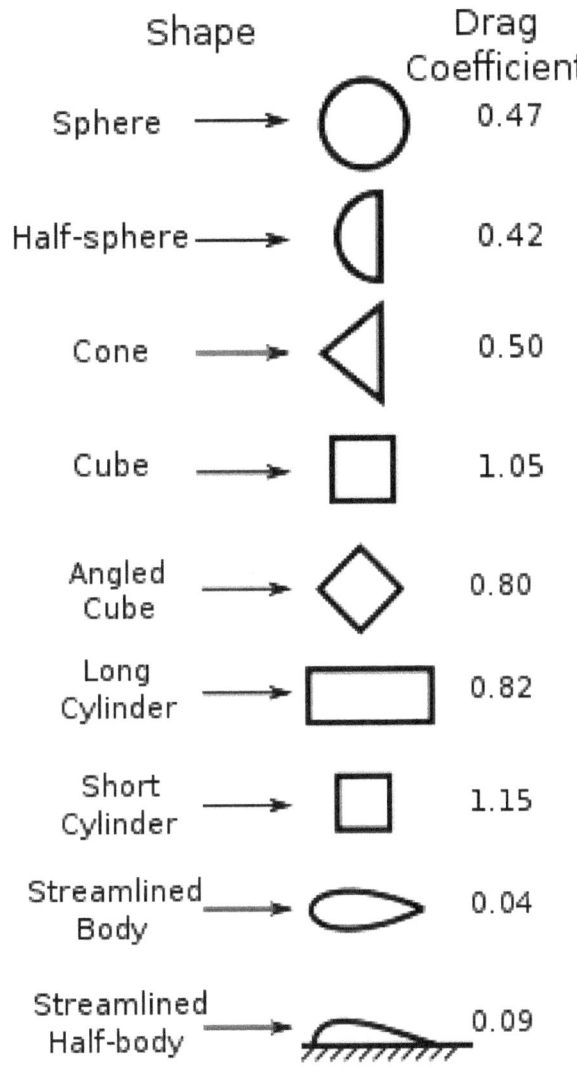

Shape	Drag Coefficient
Sphere	0.47
Half-sphere	0.42
Cone	0.50
Cube	1.05
Angled Cube	0.80
Long Cylinder	0.82
Short Cylinder	1.15
Streamlined Body	0.04
Streamlined Half-body	0.09

Measured Drag Coefficients

These tables could be extended indefinitely. For more information search on the internet for drag coefficient databases for different shapes and vehicles:

The influence of various fairings upon the drag of circular cylinders/gun barrels. From: Fig. 8.27, Fig. 8.28, Hoerner

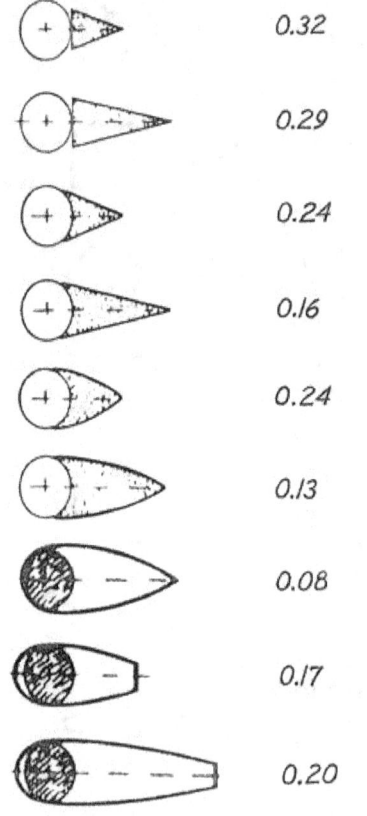

0.32

0.29

0.24

0.16

0.24

0.13

0.08

0.17

0.20

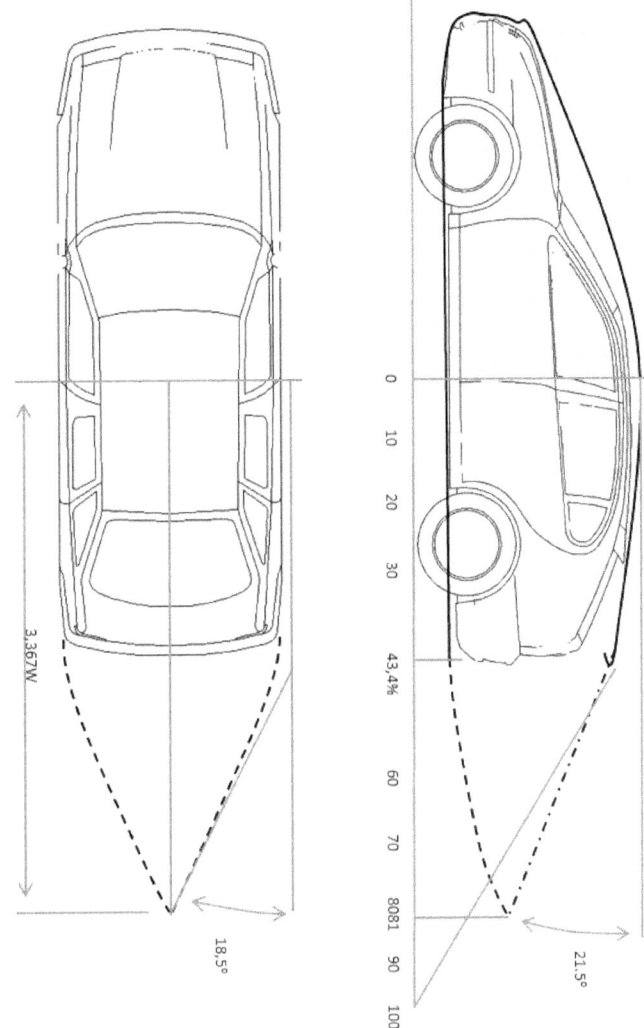

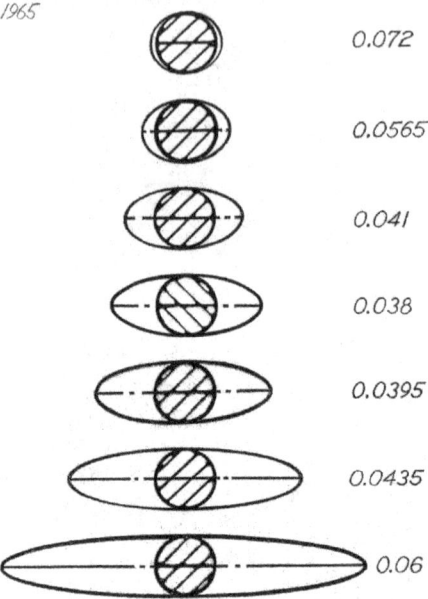

Frontal area-based
free-flight drag of Ellipsoids
as a function of fineness ratio, $R_{N} \sim 10^7$
Adapted from Hucho, 2ⁿᵈ Ed, Fig. 4.119, pg. 200 © 2013 PRK
Re.: S.F. Hoerner
1965

0.072

0.0565

0.041

0.038

0.0395

0.0435

0.06

Knowing the drag value of a car we are able to calculate the required power at a certain speed
Being "P" power, we know:
$$P = D \cdot V_{max}$$

From this expression we can derive that the relationship between power and speed is cubic;

$$P = D \cdot V = \frac{1}{2} \cdot \rho \cdot V^2 \cdot S \cdot Cd \cdot V = \frac{1}{2} \cdot \rho \cdot V^3 \cdot S \cdot Cd$$

Therefore, we can calculate the maximum speed depending on the car's power and the aerodynamic values.
Take the following example:

$$S = 2.0 \text{ m}^2$$
$$c_x = 0.32$$
$$\rho = 1.225 \text{ kg/m}^3$$

$V_1 = 80 \text{ km/h}$
$R_1 = 194 \text{ N}$
$P_1 = 4.3 \text{ kW}$

$V_2 = 160 \text{ km/h}$
$R_2 = 774 \text{ N}$
$P_2 = 34.4 \text{ kW}$

$R_2 = 4 \times R_1$

$V_2 = 2 \times V_1$

$$P_2 = R_2 \times V_2 = 4R_1 \times 2V_1 = 8\ R_1 \times V_1 = 8\ P_1$$

At double speed, we need 8 times more power.

To obtain the value of drag ("D"), we must subtract the transmission resistance values (engine, clutch, wheels, gearbox, drive shaft, etc ...).

For the top speed required power we can use the table below which depends on the type of car multiplied by the total power output by a factor to obtain the effective output:

- Single seater rear engine with narrow or cold tires (Formula Ford car mountain.): 0.91
- Single seater with wide and hot tires (Formula 3, Formula 1, Formula 3000): 0.875
- Tourism / sport car with engine in the same axis as the drive wheels (Le Mans, LMP): 0.85
- R acing cars with front engine and rear wheel drive (standard cars): 0.82

$$Force_{drag}[lbf] = \frac{\rho v^2 C_d A}{29.91}$$

$$Force_{engine}[lbf] = 375 \cdot \frac{Power[HP]}{v[mph]}$$

At top speed (v):

$$F_{drag} = F_{engine}$$

$$\frac{\rho v^2 C_d A}{29.91} = \frac{375 \cdot Power}{v}$$

The drag coefficient depends on the wings' angle of attack the Reynolds number. Let's look at an example:

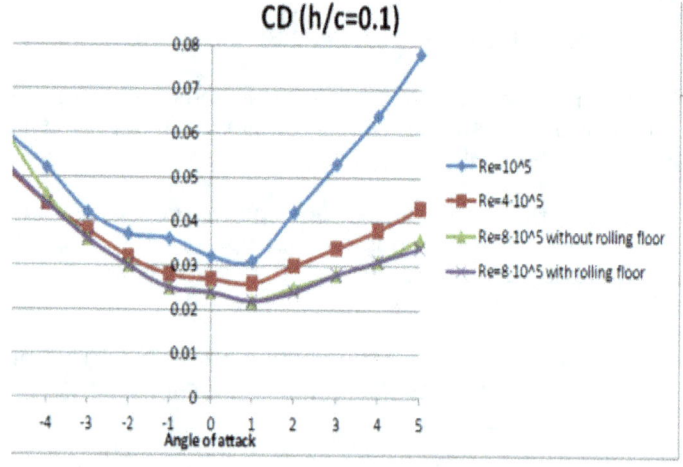

$$\rho_{air} = 0.074225 \left[\frac{lbm}{ft^3}\right] @ 75^{\circ F}$$

$$\frac{v^2 C_d A}{403} = \frac{hp \cdot 375}{v}$$

$$v^3 = \frac{hp \cdot 375 \cdot 403}{C_d A}$$

$$v^3 = \frac{hp \cdot 151130}{C_d A}$$

$$v[mph] = 53.26 \cdot \sqrt[3]{\frac{Power[hp]}{C_d A[ft^2]}}$$

We can see in the next images, tha drag coefficient in a car:

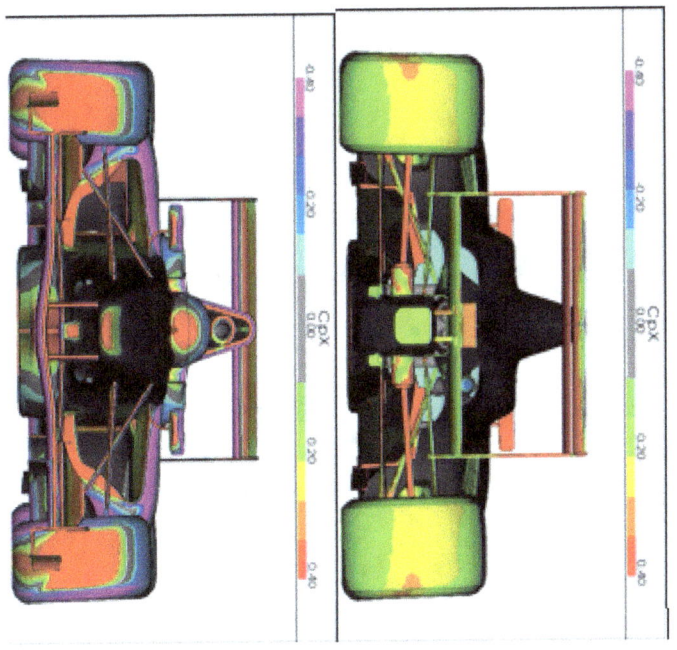

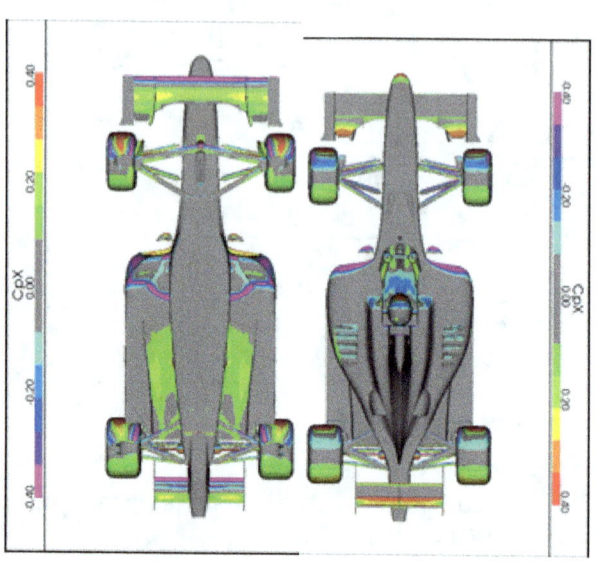

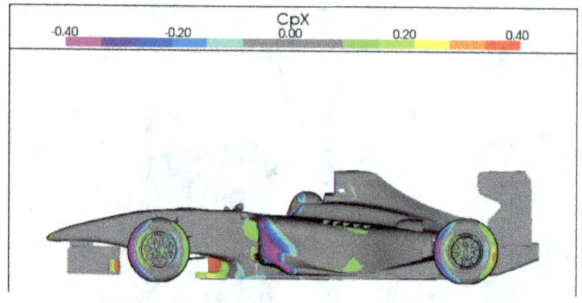

In order to reduce the consume, is necessary to reduce the aerodynamic drag; but is not the same reduce 1% the consume to all speed:

Vehicle Speed (mph)	Aerodynamic Drag Reduction to Increase Fuel Economy by 1%
60	2%
40	3%
20	6%

As we can see it also depends on the existence and size of a rolling floor in a wind tunnel:

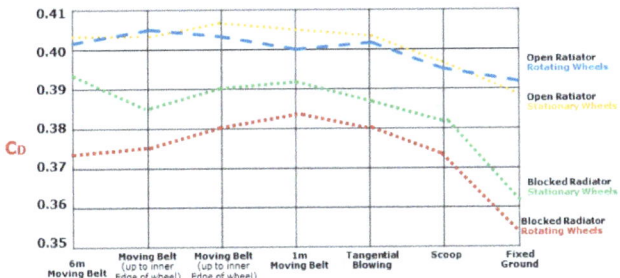

Regarding frictional resistance, we know that it can be calculated by integrating the shear stress over the entire length:

$$F_D = \int_0^L \tau_0 \, dx$$

Von Karman integral equation:

$$\tau_0 = \frac{d}{dx} \cdot \int_0^\delta \rho.u.u_\infty.dy - \frac{d}{dx}\int_0^\delta \rho.u^2.dy$$

$$\tau_0 = \frac{d}{dx} \cdot \int_0^\delta \rho . u. (u_\infty - u) dy$$

In the wall:

$$\tau_0 = \mu. \frac{\theta u}{\theta y}$$

We can now obtain the friction coefficient for both laminar and turbulent regimes:

$$\tau_0 = 0.139. \rho. u_\infty{}^2 . \frac{d\delta}{dx}$$

$$\tau_0 = \frac{7}{72} . \rho. (u_\infty)^2 . \frac{d\delta}{dx}$$

$$F_D = \frac{0.646. \rho. (u_\infty)^2 . L}{(Re_L)^{0.5}}$$

$$C_f = \frac{\tau_0}{\frac{1}{2} . \rho. (u_\infty)^2}$$

Friction coefficient

Laminar:

$$C_f = \frac{1.328}{(Re_L)^{0.5}}$$

$$C_f = \frac{0.074}{(Re_L)^{0.2}} - \frac{A}{Re_L} \quad \text{to}$$

$$5.10^5 < Re_L < 10^7$$

"A" is null, if the turbulent layer begins at the start of the surface.

Turbulent:

$$C_f = \frac{0.455}{(\log (Re_L))^{0.258}} - \frac{A}{Re_L} \quad \text{to}$$

$$Re_L > 10^7$$

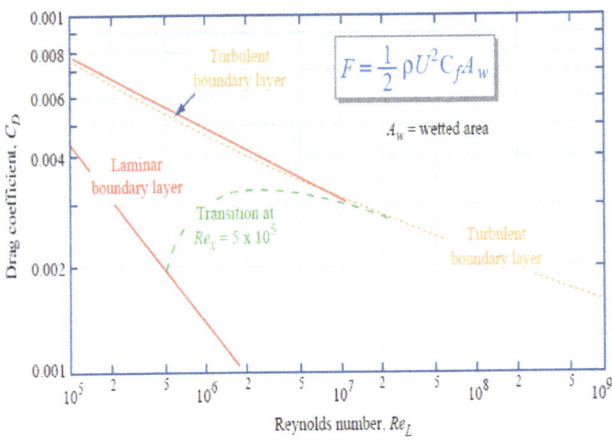

Variation of drag coefficient with Reynolds number for a smooth flat plate parallel to the flow.

To calculate the friction drag, we use the same expression applied to calculate the downforce and drag forces, but instead of using the drag or lift coefficient we use the friction coefficient.

The friction drag on F1 cars comprises up to a 15% of the total drag.

The friction drag on one hand depends on the roughness of the material, but also on the shape and size of the body; we can find bodies that have large shape drag but with little frictional drag and vice versa:

The following images illustrate this fact:

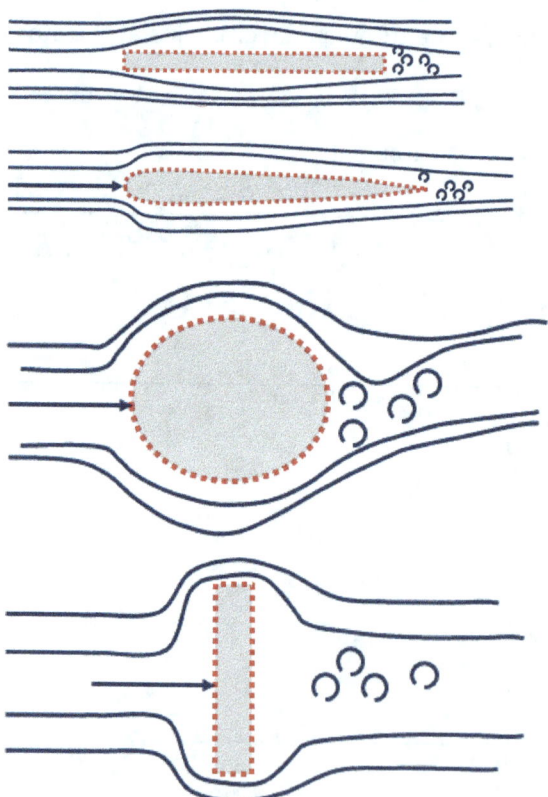

The first and second profile shapes have small drag but a lot of friction drag; the third and fourth the opposite.

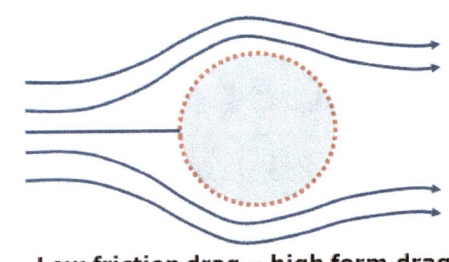

Low friction drag – high form drag

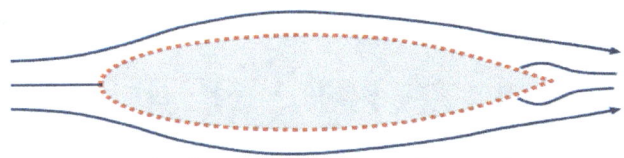

High friction drag – low form drag

The calculated expressions are mathematically quite simple, but very useful; consider the following example: suppose we have a zeppelin of 785 feet in length; 132 feet of maximum diameter; top speed 84 miles/hour; flight altitude 10,000 feet; payload of 182,000 pounds.

	Shape		C_d	
			Laminar flow	Turbulent flow
Sphere			0,47	0,27
Ellipsoid		2:1	0,27	0,06
		4:1	0,2	0,06
		8:1	0,25	0,13

We can interpolate the data from the above table so we can use it in our zeppelin: "D": C_d = 0.094.

$$Re_L = 4.59.10^8$$
$$C_f = 1.73.10^{-3}$$
$$D = 7505 \ Lb$$
$$Power = 1681 \ Horse \ Power$$

D = 17143 lb at 3840 horse power.

Actually, this zeppelin had 8 engines of 550 horsepower each, which makes a total of 4400 horses; if 3840 horses, multiplied by a value up 1.2, we get 4600 horse power, which is very close to the "real" value; this is an example of the accuracy of this type of "simple" expressions.

DRAG REDUCTION

Increasing downforce is a very important part of race car aerodynamics. However, drag reduction is also very important and should not be neglected. Why reduce drag? Basically to increase the top speed of the car and increase acceleration.

On the other hand, reducing drag means there is an important reduction in fuel consumption; see this:

Let's assume the following data for a given car:

SFC	275	g/kW·h	Specific Fuel Consumption
rho	1.225	kg/m^3	Air Density
Cd	0.67	-	Drag Coefficient
A	1.47	m^2	Frontal Area
m	920	kg/m^3	Total Mass
g	9.81	m/s^2	Gravity Acceleration
Cr	0.02	-	Friction Coefficient
Vel	120	km/h	Vehicle's Speed
transm efficiency	0.85	-	Engine Transmission Efficiency
fuel_dens	780	g/l	Fuel Density

The value of "SFC" is the fuel consumption per unit of power output.

Thus, we can first calculate the value of the aerodynamic power:

$0.5 \cdot rho \cdot Cd \cdot A \cdot ((V / 3.6) \char`^ 3) / 1000 = 22.34$ kW

Secondly we calculate the power per tire rolling on the track:

$Cr \cdot m \cdot g \cdot (V / 3.6) / 1000 = 6.01$ kW

Which makes a sum of 28.36 kW = Total Power

Now, we calculate the fuel consumption:

$SFC \cdot PotTotal \cdot (100 / V) / f\_dens = 8.33$ liters / 100 km

With this template we can obtain some important conclusions: Reducing a car's weight will not reduce significantly its fuel consumption because mass only affects tire rolling power not aerodynamic power which generates a far greater power. When we study a vehicle with the intention to reduce its fuel consumption we must reduce aerodynamic drag. One way could be to think of a way to generate positive lift (without being dangerous) as this lift will reduce the frictional drag.

Let's see how to reduce frictional drag. There are several methods:

- Make the surface smoother to reduce friction drag;

This is a generic method which aims to reduce the frictional resistance by generating a turbulent boundary layer on the majority of the surface; this entails a reduction in friction because there is less air in contact with the surface. The transition to a turbulent boundary layer is made through vortex generators or turbulators:

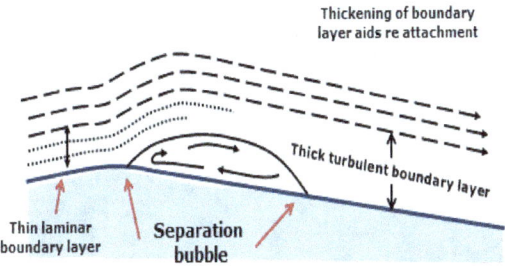

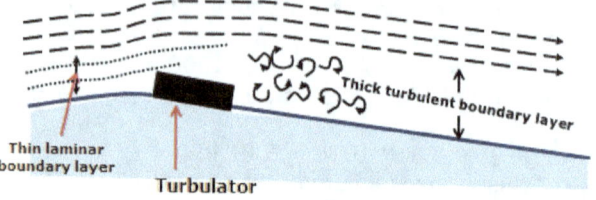

Thin laminar boundary layer

Turbulator

Thick turbulent boundary layer

- Another method similar to the one above is to use a technique to reduce the amount of air close to the surface and with a lower density; for this we could use the geometry seen in the picture below. As we can see, it has a rough skin which is able to create that characteristic layer of air:

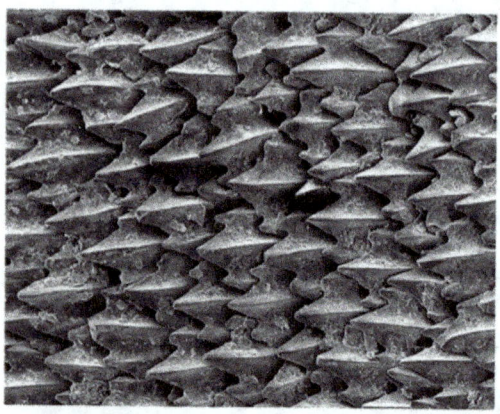

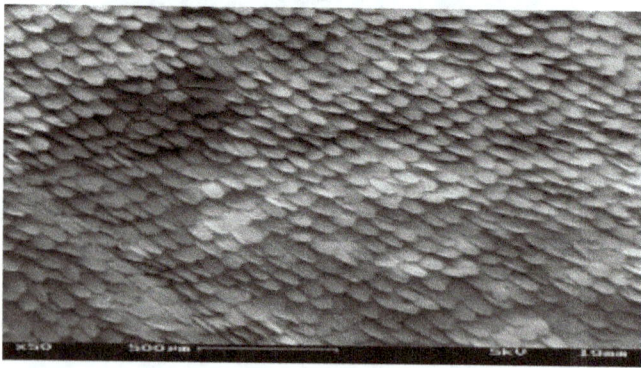

You may do the same, by adding dimples on the surface; the existence of these holes, "can cause" a reduction in drag by reducing friction:

Remembering the Magnus effect, this was "valid" only when the body rotated. But we can strategically place dimples on a surface and reduce drag. This is often done with bicycle wheels or in a racing bicycle helmets:

- Another method: It is rarer and used exclusively, by the moment, in water. This method tries to reduce the density of water in contact with the surface, causing what is known as cavitation. The idea behind the concept is that the body is encapsulated under a fluid with low pressure and density. In the case of torpedoes, the water is transformed into steam, the friction between the body and the steam is much lower than between the body and liquid water; thus it is able to reach supersonic speeds in water. The problem resides in how we manage to get this effect (This is a drag reduction method very "appreciated" for military purposes...)

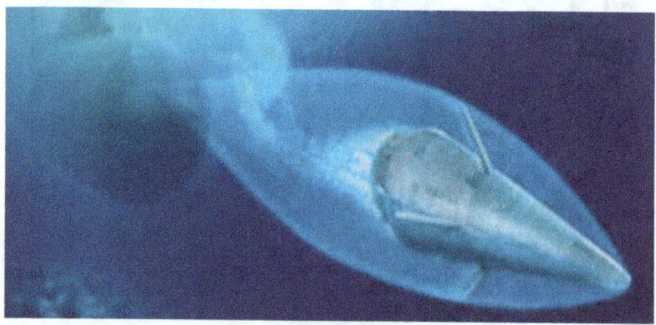

Wikipedia.

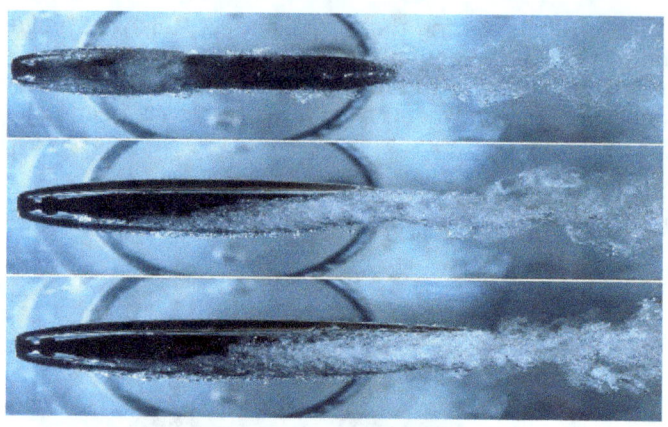

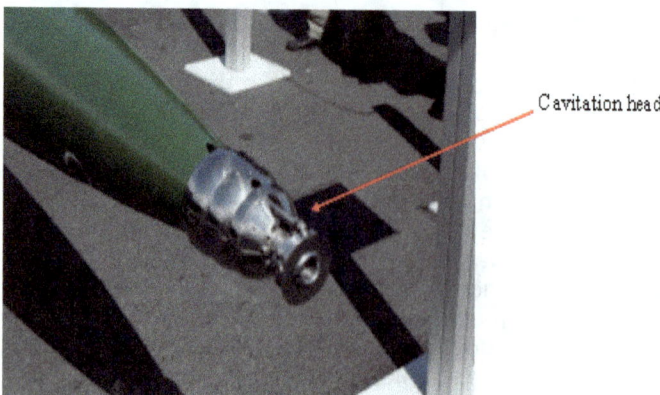

Cavitation head

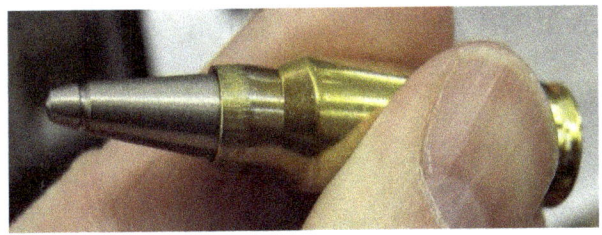

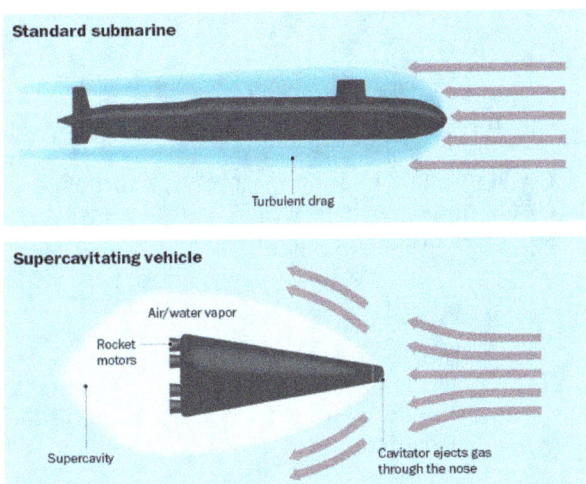

Standard submarine

Turbulent drag

Supercavitating vehicle

Air/water vapor

Rocket motors

Supercavity

Cavitator ejects gas through the nose

Exist a project of Citroen, named Supercav model; that, inject air forward, from front car:

SUPERCAVITATION SCIENCE THEORY

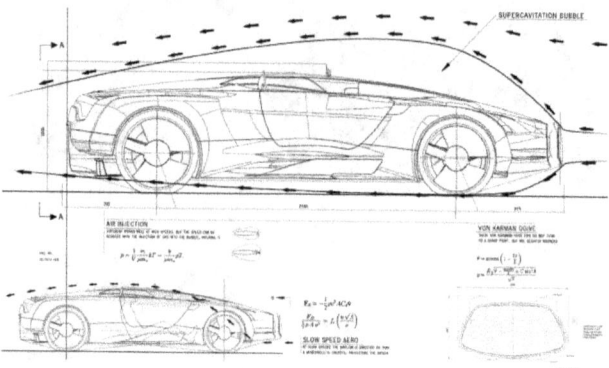

One method about, as a resume of this last, is:

It will possible heating the object, in order to create above the surfaces, lower air density, and so, lower drag... genial, no ????

- The following method is radically different and is based on modifying the geometry of the body to reduce drag in general, not just friction.

In aerodynamics, there are many things that are not what they seem; for example, the following two figures, circle and profile, are drawn to scale, and both have the same air resistance:

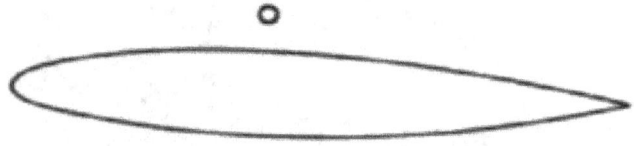

The vehicle's shape doesn't consist only of the frontal area.

Look at the following images: We see a bus, with different geometries to reduce drag. We can see a considerable reduction in design (3) due to softened edges as compared to design (1).

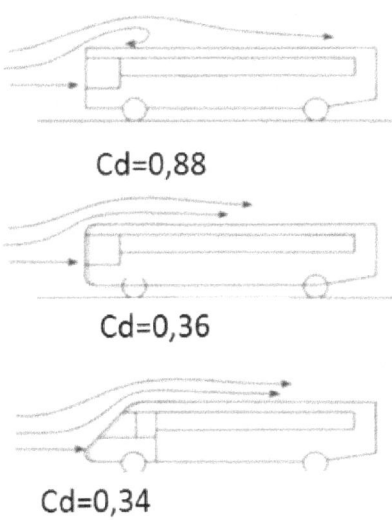

Cd=0,88

Cd=0,36

Cd=0,34

The details, are very specials: as example: SCIROCCO VOLSKWAGEN:

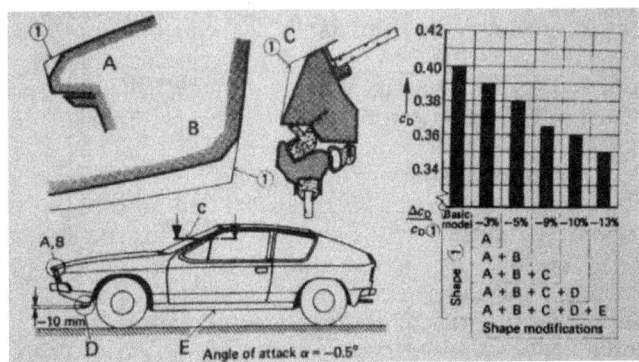

So far, we have emphasized that by softening the frontal shape by rounding the edges we could reduce drag significantly. If we do the same with the rear the consequences are not the same:

Suppose we have a truck, with 3 different rear designs:

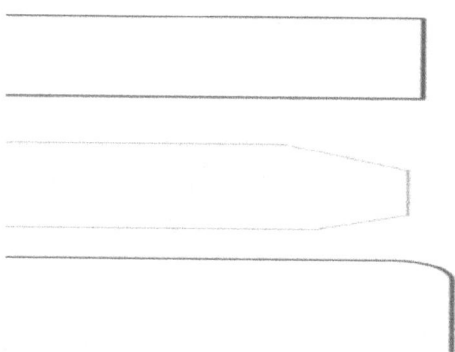

The cutting angle for the second design is 15 °; and for the third design there are 2 alternatives, one with a rounding radius of 10 cm and the other with a radius of 20 cm; let's take a look at the results of the study:

Base: Cd = 0.325
15°: Cd = 0.23
10 cm: Cd = 0.293
20 cm: Cd = 0.28

A reduction of around 30% is achieved by cutting the rear (design 2). Rounding the edges as we did in the front previously only manages a reduction of almost 12%; the difference is very noticeable. Essentially, to reduce drag we try to create a body whose motion doesn't interfere or modify the air at all.

This is absolutely impossible, but it is important that we reduce drag as much as we can so that we can succeed in reducing this influence of the car on the air; i.e. make the air similar downstream to how it was previous to hitting the car.

There are a number of aerodynamic principles, you should know and follow:

• Blunt or rounded front and rear pointed

bodies.
- Eliminate turbulence.
- Avoid gaps between surfaces.
- Avoid low pressures in general.

Under these simple assumptions, we are able to reduce drag of any vehicle and provide it with greater speed with less power consumption.

Avoiding gaps involves filling areas of low pressure; we will work on a truck to optimize its drag (Wikipedia images):

In general, the contribution of each part of the truck to the total resistance is:

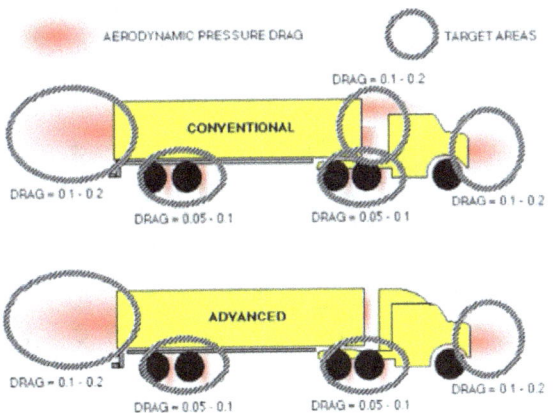

We will incorporate systems to reduce the overall aerodynamic resistance.

a) Front:

In the front we could try to make a rounded cabin:

However, the system which is proliferating in the roads is the one that is placed on top of the cabin. Deflecting the flow preventing the air from hitting directly the trailer:

Figure 13 Common Aerodynamic Features

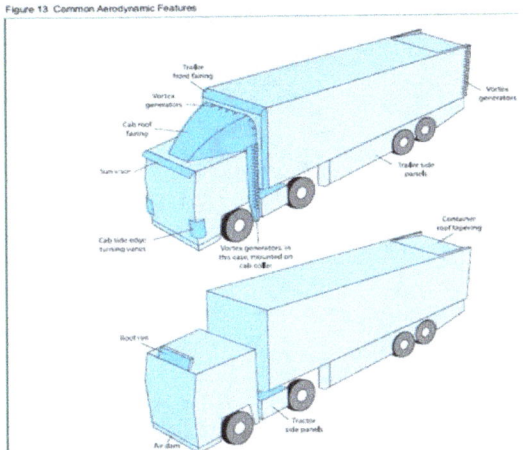

b) Fill the the gap left between the cabin and the trailer:

We can also do a combination of the 2 methods seen up to now:

We could design this area between as a circular ring, so that it allows rotation:

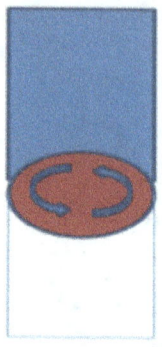

c) Side skirts:

It is based on the same principle as the previous systems: The goal is to try to remove the gap left between the wheels and the sides; closing this gap involves reducing drag significantly.

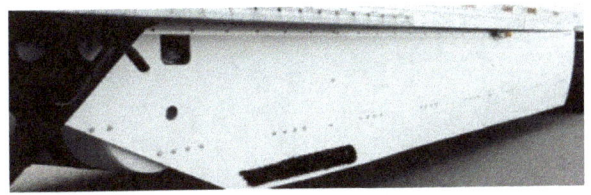

d) Filling the rear:

It is the most important method in terms of amount of drag reduction. It is based on the filling of the depression that is generated in the rear of any vehicle: A car, moving or moved, leaves a space with less air than there was previously. We usually call this a wake. Depending on the geometry of the car, this "gap" is bigger or smaller; filling quickly the vacuum is a directly related with the aerodynamic drag.

This filling can be done in many ways and through many systems:

- Taking air directly from the top, and introducing it through channels in the low pressure zone, simple and easy:

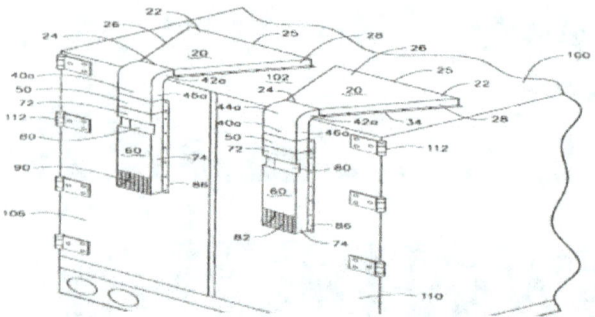

This type of system is called self-generating; The higher the speed the truck has, the greater the depression in the rear and therefore higher will be the amount of air sucked from the top; It is a system that regulates itself somehow; whenever we connect an area of high and low pressure, we obtain a self-generating system.

- Turbulators:

We already know the tendency of the turbulent regime to adhere more and the ability to achieve higher angles of attack; For this reason, we can force a turbulent regime at the end of the trailer:

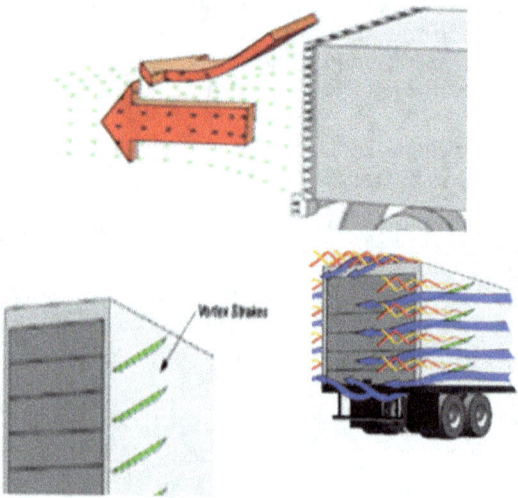

About that, in race cars, in particular in rear wing, there is somethink like that, with the same function:

www.ingramcontent.com/pod-product-compliance
Lightning Source LLC
Chambersburg PA
CBHW071135220526
45467CB00015B/1049